Explanatory Model Analysis

CHAPMAN & HALL/CRC DATA SCIENCE SERIES

Reflecting the interdisciplinary nature of the field, this book series brings together researchers, practitioners, and instructors from statistics, computer science, machine learning, and analytics. The series will publish cutting-edge research, industry applications, and textbooks in data science.

The inclusion of concrete examples, applications, and methods is highly encouraged. The scope of the series includes titles in the areas of machine learning, pattern recognition, predictive analytics, business analytics, Big Data, visualization, programming, software, learning analytics, data wrangling, interactive graphics, and reproducible research.

Published Titles

Feature Engineering and Selection
A Practical Approach for Predictive Models
Max Kuhn and Kjell Johnson

Probability and Statistics for Data Science
Math + R + Data
Norman Matloff

Introduction to Data Science
Data Analysis and Prediction Algorithms with R
Rafael A. Irizarry

Cybersecurity Analytics
Rakesh M. Verma and David J. Marchette

Basketball Data Science
With Applications in R
Paola Zuccolotto and Marcia Manisera

JavaScript for Data Science
Maya Gans, Toby Hodges, and Greg Wilson

Statistical Foundations of Data Science
Jianqing Fan, Runze Li, Cun-Hui Zhang, and Hui Zou

Explanatory Model Analysis
Explore, Explain, and Examine Predictive Models
Przemyslaw Biecek, Tomasz Burzykowski

For more information about this series, please visit: https://www.crcpress.com/Chapman--HallCRC-Data-Science-Series/book-series/CHDSS

Explanatory Model Analysis

Explore, Explain and Examine Predictive Models

Przemyslaw Biecek
Tomasz Burzykowski

CRC Press
Taylor & Francis Group
Boca Raton London New York

CRC Press is an imprint of the
Taylor & Francis Group, an **informa** business

A CHAPMAN & HALL BOOK

First edition published 2021
by CRC Press
6000 Broken Sound Parkway NW, Suite 300, Boca Raton, FL 33487-2742

and by CRC Press
2 Park Square, Milton Park, Abingdon, Oxon, OX14 4RN
© 2021 Taylor & Francis Group, LLC

ISBN: 9780367135591 (hbk)
ISBN: 9780367693923 (pbk)
ISBN: 9780429027192 (ebk)

Typeset in Minion
by KnowledgeWorks Global Ltd.

Małgorzacie Henrykowi
Agnieszce

P.B.

Mojej Mamie, Mikołajowi i Beacie

T.B.

Contents

Part I

Introduction

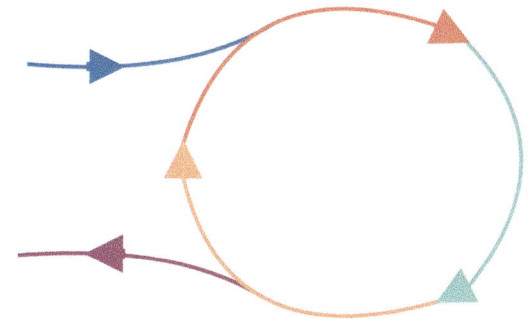

1

Introduction

1.1 The aim of the book

Predictive models are used to guess (statisticians would say: predict) values of a variable of interest based on values of other variables. As an example, consider prediction of sales based on historical data, prediction of the risk of heart disease based on patient's characteristics, or prediction of political attitudes based on Facebook comments.

Predictive models have been used throughout the entire human history. Ancient Egyptians, for instance, used observations of the rising of Sirius to predict the flooding of the Nile. A more rigorous approach to model construction may be attributed to the method of least squares, published more than two centuries ago by Legendre in 1805 and by Gauss in 1809. With time, the number of applications in the economics, medicine, biology, and agriculture has grown. The term *regression* was coined by Francis Galton in 1886. Initially, it referred to biological applications, while today it is used for various models that allow prediction of continuous variables. Prediction of nominal variables is called *classification*, and its beginning may be attributed to works of Ronald Fisher in 1936.

During the last century, many statistical models that can be used for predictive purposes have been developed. These include linear models, generalized linear models, classification and regression trees, rule-based models, and many others. Developments in mathematical foundations of predictive models were boosted by increasing computational power of personal computers and availability of large datasets in the era of "big data" that we have entered.

With the increasing demand for predictive models, model properties such as flexibility, capability of internal variable selection or feature engineering, and high precision of predictions are of interest. To obtain robust models, ensembles of models are used. Techniques like bagging, boosting, or model stacking combine hundreds or thousands of simpler models into one super-model. Large deep-neural models may have over a billion parameters.

There is a cost of this progress. Complex models may seem to operate like "black boxes". It may be difficult, or even impossible, to understand how thousands of variables affect a model's prediction. At the same time, complex models may not work as well as we would like them to. An overview of real problems with massive-scale black-box models may be found in an excellent book by O'Neil (2016) or in her TED Talk "The era of blind faith in big data must end". There is a growing number of examples of predictive models with performance that deteriorated over time or became biased in some sense. For instance, IBM's Watson for Oncology was criticized by oncologists for delivering unsafe and inaccurate recommendations (Ross and Swetliz, 2018). Amazon's system for curriculum-vitae screening was found to be biased against women (Dastin, 2018). The COMPAS (Correctional Offender Management Profiling for Alternative Sanctions) algorithm for predicting recidivism, developed by Northpointe (now Equivant), was accused of bias against African-Americans (Larson et al., 2016). Algorithms behind the Apple Credit Card are accused of being gender-biased (Duffy, 2019). Some tools for sentiment analysis are suspected of age-bias (Diaz et al., 2018). These are examples of models and algorithms that led to serious violations of fairness and ethical principles. An example of a situation when data-drift led to a deterioration in model performance is the Google Flu model, which gave worse predictions after two years than at baseline (Salzberg, 2014; Lazer et al., 2014).

A reaction to some of these examples and issues are new regulations, like the General Data Protection Regulation (GDPR, 2018). Also, new civic rights are being formulated (Goodman and Flaxman, 2017; Casey et al., 2019; Ruiz, 2018). A noteworthy example is the "Right to Explanation", i.e., the right to be provided with an explanation for an output of an automated algorithm (Goodman and Flaxman, 2017). To exercise the right, we need new methods for verification, exploration, and explanation of predictive models.

Figure 1.1 presents an attempt to summarize how the increase in the model complexity affects the relative importance of domain understanding, the choice of a model, and model validation.

In classical statistics, models are often built as a result of a good understanding of the application domain. Domain knowledge helps to create and select the most important variables that can be included in relatively simple models that yield predictive scores. Model validation is based mainly on the evaluation of the goodness-of-fit and hypothesis testing. Statistical hypotheses shall be stated before data analysis, and obtained p-values should not interfere with the way in which data were processed or models were constructed.

Machine learning, on the other hand, exploits the trade-off between the availability of data and domain knowledge. The effort is shifted from a deep understanding of the application domain towards (computationally heavy) construction and fitting of models. Flexible models can use massive amounts of data to select informative variables and filter out uninformative ones. The validation step gains in importance because it provides feedback to the model construction.

How might this approach look in the future? It is possible that the increasing automation of the exploratory data analysis (EDA) and the modelling part of the process will shift the focus towards the validation of a model. In particular, validation will not only focus on how good a model's fit and predictions are but also what other risks (like concept drift) or biases may be associated with the model. Model exploration will allow us to better and faster understand the analyzed data.

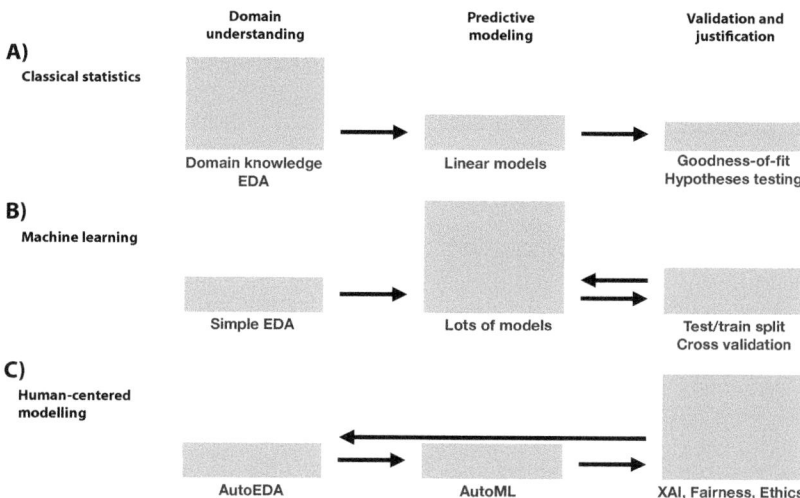

FIGURE 1.1 Shift in the relative importance and effort (symbolically represented by the shaded boxes) put in different phases of data-driven modelling. Arrows show feedback loops in the modelling process. (A) In classical statistics, modelling is often based on a deep understanding of the application domain combined with exploratory data analysis (EDA). Most often, (generalized) linear models are used. Model validation includes goodness-of-fit evaluation and hypothesis testing. (B) In machine learning (ML), domain knowledge and EDA are often limited. Instead, flexible models are fitted to large volumes of data to obtain a model offering a good predictive performance. Evaluation of the performance (applying strategies like cross-validation to deal with overfitting) gains in importance, as validation provides feedback to model construction. (C) In the (near?) future, auto-EDA and auto-ML will shift focus even further to model validation that will include the use of explainable artificial intelligence (XAI) techniques and evaluation of fairness, ethics, etc. The feedback loop is even longer now, as the results from model validation will also be helping in domain understanding.

Summarizing, we can conclude that, today, the true bottleneck in predictive modelling is neither the lack of data, nor the lack of computational power, nor inadequate algorithms, nor the lack of flexible models. It is the lack of tools for model *exploration* and, in particular, model *explanation* (obtaining insight into model-based predictions) and model *examination* (evaluation of model's performance and understanding the weaknesses). Thus, in this book, we present a collection of methods that may be used for this purpose. As development of such methods is a very active area of research, with new methods becoming available almost on a continuous basis, we do not aim at being exhaustive. Rather, we present the mindset, key concepts and issues, and several examples of methods that can be used in model exploration.

1.2   A bit of philosophy: three laws of model explanation

In 1942, in his story "Runaround", Isaac Asimov formulated *Three Laws of Robotics*:

1)   a robot may not injure a human being,
2)   a robot must obey the orders given it by human beings, and
3)   a robot must protect its own existence.

Today's robots, like cleaning robots, robotic pets, or autonomous cars are far from being conscious enough to fall under Asimov's ethics. However, we are more and more surrounded by complex predictive models and algorithms used for decision-making. Artificial-intelligence models are used in health care, politics, education, justice, and many other areas. The models and algorithms have a far larger influence on our lives than physical robots. Yet, applications of such models are left unregulated despite examples of their potential harmfulness. An excellent overview of selected issues is offered in the book by O'Neil (2016).

It is now becoming clear that we have got to control the models and algorithms that may affect us. Asimov's laws are being referred to in the context of the discussion around *ethics of artificial intelligence* (https://en.wikipedia.org/wiki/Ethics_of_artificial_intelligence). Initiatives to formulate principles for artificial-intelligence development have been undertaken, for instance, in the UK (Olhede and Wolfe, 2018). Following Asimov's approach, we propose three requirements that any predictive model should fulfil:

- **Prediction's validation.** For every prediction of a model, one should be able to verify how strong the evidence is that supports the prediction.
- **Prediction's justification.** For every prediction of a model, one should be able to understand which variables affect the prediction and to what extent.
- **Prediction's speculation.** For every prediction of a model, one should be

able to understand how the prediction would change if the values of the variables included in the model changed.

We see two ways to comply with these requirements. One is to use only models that fulfil these conditions by design. These are so-called "interpretable-by-design models" that include linear models, rule-based models, or classification trees with a small number of parameters (Molnar, 2019). However, the price of transparency may be a reduction in performance. Another way is to use tools that allow, perhaps by using approximations or simplifications, "explaining" predictions for any model. In our book, we focus on the latter approach.

1.3   Terminology

It is worth noting that, when it comes to predictive models, the same concepts have often been given different names in statistics and in machine learning. In his famous article (Breiman, 2001b), Leo Breiman described similarities and differences in perspectives used by the two communities. For instance, in the statistical-modelling literature, one refers to "explanatory variables", with "independent variables", "predictors", or "covariates" often used as equivalents. Explanatory variables are used in a model as a means to explain (predict) the "dependent variable", also called "predicted" variable or "response". In machine-learning terminology, "input variables" or "features" are used to predict the "output" or "target" variable. In statistical modelling, models are "fit" to the data that contain "observations", whereas in the machine-learning world a model is "trained" on a dataset that may contain "instances" or "cases". When we talk about numerical constants that define a particular version of a model, in statistical modelling, we refer to model "coefficients", while in machine learning it is more customary to refer to model "parameters". In statistics, it is common to say that model coefficients are "estimated", while in machine learning it is more common to say that parameters are "trained".

To the extent possible, in our book we try to consistently use the statistical-modelling terminology. However, the reader may find references to a "feature" here and there. Somewhat inconsistently, we also introduce the term "instance-level" explanation. Instance-level explanation methods are designed to extract information about the behaviour of a model related to a specific observation (or instance). On the other hand, "dataset-level" explanation techniques allow obtaining information about the behaviour of the model for an entire dataset.

We consider models for dependent variables that can be continuous or categorical. The values of a continuous variable can be represented by numbers with an ordering that makes some sense (ZIP-codes or phone numbers are not considered as continuous variables, while age or number of children are).

A continuous variable does not have to be continuous in the mathematical sense; counts (number of floors, steps, etc.) will be treated as continuous variables as well. A categorical variable can assume only a finite set of values that are not numbers in the mathematical sense, i.e., it makes no sense to subtract or divide these values.

In this book, we treat models as "black boxes". We don't assume anything about their internal structure or complexity. We discuss the specificity of such an approach in a bit more detail in the next section.

1.4   Black-box models and glass-box models

Usually, the term "black-box" model is used for models with a complex structure that is hard to understand by humans. This usually refers to a large number of model coefficients or complex mathematical transformations. As people vary in their capacity to understand complex models, there is no strict threshold for the number of coefficients that makes a model a black box. In practice, for most people, this threshold is probably closer to 10 than to 100.

A "glass-box" (sometimes also called a "white-box" or a "transparent-box") model, which is opposite to a black-box one, is a model that is easy to understand (though maybe not by every person). It has a simple structure and a limited number of coefficients.

The most common classes of glass-box models are decision or regression trees (see an example in Figure 1.2), or models with an explicit compact structure. As an example of the latter, consider a model for obesity based on the body-mass index (BMI), with BMI defined as the mass (in kilograms) divided by the square of height (in meters). Subjects are classified as *underweight* if their BMI<18, as *normal* if their BMI lies in the interval [18,25], and as *overweight* if their BMI>25. The compact form of the model makes it easy to understand, for example, how does a change in BMI change the predicted obesity class.

The structure of a glass-box model is, in general, easy to understand. It may be difficult to collect the necessary data, build the model, fit it to the data, or perform model validation, but once the model has been developed its interpretation and mode of working is straightforward.

Why is it important to understand a model's structure? There are several important advantages. If the structure is transparent, we can easily see which explanatory variables are included in the model and which are not. Hence, for instance, we may be able to question the model from which a particular explanatory variable is excluded. Also, in the case of a model with a transparent structure and a limited number of coefficients, we can easily link changes in the model's predictions with changes in particular explanatory variables. This, in turn, may allow us to challenge the model on the ground of domain knowledge if, for instance, the effect of a particular variable on predictions is inconsistent with previously-established results. Note that linking changes in the model's predictions to changes in particular explanatory variables may be difficult when there are many variables and/or coefficients in the model. For instance, a classification tree with hundreds of nodes is difficult to understand, as is a linear regression model with hundreds of coefficients.

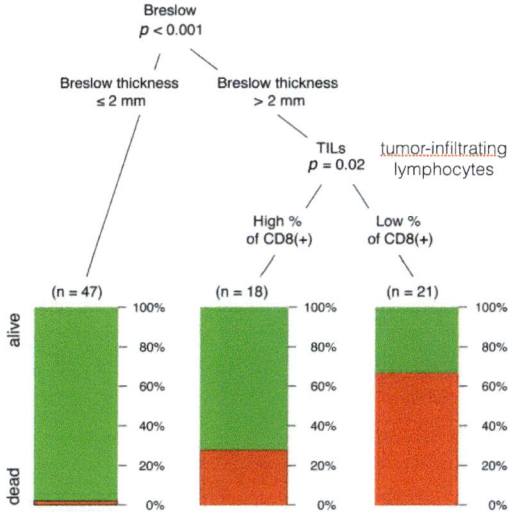

FIGURE 1.2 An example of a decision-tree model for melanoma risk patients developed by Donizy et al. (2016). The model is based on two explanatory variables, Breslow thickness and the presence of tumor infiltration lymphocytes. These two variables classify patients into three groups with a different probability of survival.

Note that some glass-box models, like the decision-tree model presented in Figure 1.2, satisfy by design the explainability laws introduced in Section 1.2. In particular, regarding *prediction's validation*, we see how many patients fall in a given category in each node. With respect to *prediction's justification*, we know which explanatory variables are used in every decision path. Finally, regarding *prediction's speculation*, we can trace how changes in particular variables will affect the model's prediction. We can, of course, argue if the model is good or not, but the model structure is obviously transparent.

Comprehending the performance of black-box models presents more challenges. The structure of a complex model, such as, for example, a neural-network model, may be far from transparent. Consequently, we may not understand which features influence the model decisions and by how much. Consequently, it may be difficult to decide whether the model is consistent with our domain knowledge.

In our book, we present tools that can help in extracting the information necessary for the evaluation of models in a model-agnostic fashion, i.e., in the same way regardless of the complexity of the analyzed model.

1.5 Model-agnostic and model-specific approach

Interest in model interpretability is as old as statistical modelling itself. Some classes of models have been developed for a long period or have attracted intensive research. Consequently, those classes of models are equipped with excellent tools for model exploration, validation, or visualisation. For example: /index%7BModel-agnostic approach}

- There are many tools for diagnostics and evaluation of linear models (see, for example, Galecki and Burzykowski (2013) or Faraway (2005)). Model assumptions are formally defined (normality, linear structure, homogeneity of variance) and can be checked by using normality tests or plots (like normal qq-plots), diagnostic plots, tests for model structure, tools for identification of outliers, etc. A similar situation applies to generalized linear models (see, for example, Dobson (2002)).
- For more advanced models with an additive structure, like the proportional hazards model, many tools can be used for checking model assumptions (see, for example, Harrell Jr (2018) or Sheather (2009)).
- Random forest models are equipped with the out-of-bag method of evaluating performance and several tools for measuring variable importance (Breiman et al., 2018). Methods have been developed to extract information about possible interactions from the model structure (Paluszynska and Biecek, 2017; Ehrlinger, 2016). Similar tools have been developed for other ensembles of trees, like boosting models (see, for example, Foster (2017) or Karbowiak and Biecek (2019)).
- Neural networks enjoy a large collection of dedicated model-explanation tools that use, for instance, the layer-wise relevance propagation technique (Bach et al., 2015), saliency maps technique (Simonyan et al., 2014), or a mixed approach. A summary can be found in Samek et al. (2018) and Alber et al. (2019).
- The Bidirectional Encoder Representations from Transformers (BERT) family of models leads to high-performance models in Natural Language Processing. The exBERT method (Hoover et al., 2020) is designed to visualize the activation of attention heads in this model.

Of course, the list of model classes with dedicated collections of model-explanation and/or diagnostics methods is much longer. This variety of model-specific approaches does lead to issues, though. For instance, one cannot easily compare explanations for two models with different structures. Also, every time a new architecture or a new ensemble of models is proposed, one needs to look for new methods of model exploration. Finally, no tools for model explanation or diagnostics may be immediately available for brand-new models.

For these reasons, in our book we focus on model-agnostic techniques. In particular, we prefer not to assume anything about the model structure, as we may be dealing with a black-box model with an unspecified structure. Note that often we do not have access to model coefficients, but only to a specified Application Programming Interface (API) that allows querying remote models as, for example, in Microsoft Cognitive Services (Azure, 2019). In that case, the only operation that we may be able to perform is the evaluation of a model on a specified set of data.

However, while we do not assume anything about the structure of the model, we will assume that the model operates on p-dimensional vector of explanatory variables/features and, for a single observation, it returns a single value (score/probability), which is a real number. This assumption holds for a broad range of models for data such as tabular data, images, text data, videos, etc. It may not be suitable for, e.g., models with memory-like sequence-to-sequence models (Sutskever et al., 2014) or Long Short-Term Memory models (Hochreiter and Schmidhuber, 1997) in which the model output depends also on sequence of previous inputs, or generative models that output text of images.

1.6 The structure of the book

This book is split into four major parts. In the first part, *Introduction*, we introduce notation, datasets, and models used in the book. In the second part, *Instance-level Exploration*, we present techniques for exploration and explanation of a model's predictions for a single observation. In the third part, *Dataset-level Exploration*, we present techniques for exploration and explanation of a model for an entire dataset. In the fourth part, *Use-case*, we apply the methods presented in the previous parts to an example in which we want to assess the value of a football player. The structure of the second and the third part is presented in Figure 1.3.

In more detail, the first part of the book consists of Chapters 2–4. In Chapter 2, we provide a short introduction to the process of data exploration and model construction, together with notation and definition of key concepts that are used in consecutive chapters. Moreover, in Chapters 3.1 and 3.2, we provide a short description of R and Python tools and packages that are necessary to replicate the results presented in the book. Finally, in Chapter 4, we describe two datasets that are used throughout the book to illustrate the presented methods and tools.

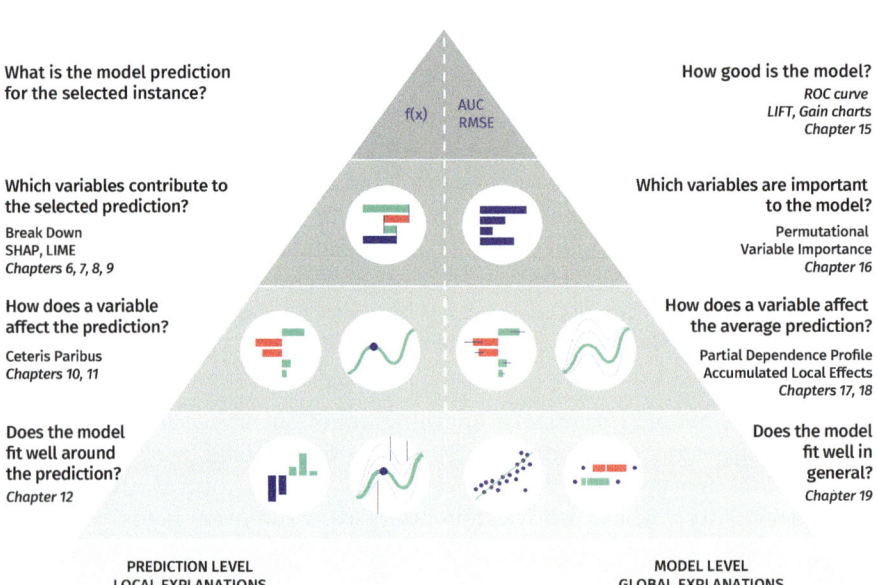

FIGURE 1.3 Model exploration methods presented in the book. The left-hand side (corresponding to the second part of the book) focuses on instance-level exploration, while the right-hand side (corresponding to the third part of the book) focuses on dataset-level exploration. Consecutive layers of the stack are linked with a deeper level of model exploration. The layers are linked with laws of model exploration introduced in Section 1.2.

The second part of the book focuses on instance-level explainers and consists of Chapters 6–13. Chapters 6–8 present methods that allow decomposing a model's predictions into contributions corresponding to each explanatory variable. In particular, Chapter 6 introduces break-down (BD) for additive attributions for predictive models, while Chapter 7 extends this method to attributions that include interactions. Chapter 8 describes Shapley Additive Explanations (SHAP) (Lundberg and Lee, 2017), an alternative method for decomposing a model's predictions that is closely linked with Shapley values developed originally for cooperative games by Shapley (1953). Chapter 9 presents a different approach to the explanation of single-instance predictions. It is based on a local approximation of a black-box model by a simpler glass-box one. In this chapter, we discuss the Local-Interpretable Model-agnostic Explanations (LIME) method (Ribeiro et al., 2016). These chapters correspond to the second layer of the stack presented in Figure 1.3.

In Chapters 10–12 we present methods based on the ceteris-paribus (CP) profiles. The profiles show the change of model-based predictions induced by a change of a single explanatory-variable. The profiles are introduced in Chapter 10, while Chapter 11 presents a CP-profile-based measure that summarizes the impact of a selected variable on the model's predictions. The measure can be used to determine the order of variables in model exploration. It is particularly important for models with large numbers of explanatory variables. Chapter 12 focuses on model diagnostics. It describes local-stability plots that are useful to investigate the sources of a poor prediction for a particular single observation.

The final chapter of the second part, Chapter 13, compares various methods of instance-level exploration.

The third part of the book focuses on dataset-level exploration and consists of Chapters 14–19. The chapters present methods in the same order as shown in the right-hand side of Figure 1.3. In particular, Chapter 15 presents measures that are useful for the evaluation of the overall performance of a predictive model. Chapter 16 describes methods that are useful for the evaluation of an explanatory-variable's importance. Chapters 17 and 18 introduce partial-dependence and accumulated-dependence methods for univariate exploration of a variable's effect. These methods correspond to the third (from the top) layer of the right-hand side of the stack presented in Figure 1.3. The final chapter of this part of the book is Chapter 19 that summarises diagnostic techniques based on model residuals.

The book is concluded with Chapter 21 that presents a worked-out example of model-development process in which we apply all the methods discussed in the second and third part of the book.

To make the exploration of the book easier, each chapter of the second and the third part of the book has the same structure:

- Section *Introduction* explains the goal of the method(s) presented in the chapter.
- Section *Intuition* explains the general idea underlying the construction of the method(s) presented in the chapter.
- Section *Method* shows mathematical details related to the method(s). This section can be skipped if you are not interested in the details.
- Section *Example* shows an exemplary application of the method(s) with discussion of results.
- Section *Pros and cons* summarizes the advantages and disadvantages of the method(s). It also provides some guidance regarding when to use the method(s).
- Section *Code snippets* shows the implementation of the method(s) in R and Python. This subsection can be skipped if you are not interested in the implementation.

1.7 What is included in this book and what is not

The area of model exploration and explainability is quickly growing and is present in many different flavors. Instead of showing every existing method (is it really possible?), we rather selected a subset of consistent tools that form a good starting toolbox for model exploration. We mainly focus on the impact of the model exploration and explanation tools rather than on selected methods. We believe that by providing the knowledge about the potential of model exploration methods and about the language of model explanation, we will help the reader in improving the process of data modelling.

Taking this goal into account **in this book, we do show**

- how to determine which explanatory variables affect a model's prediction for a single observation. In particular, we present the theory and examples of methods that can be used to explain prediction like break-down plots, ceteris-paribus profiles, local-model approximations, or Shapley values;
- techniques to examine predictive models as a whole. In particular, we review the theory and examples of methods that can be used to explain model performance globally, like partial-dependence plots or variable-importance plots;
- charts that can be used to present the key information in a quick way;
- tools and methods for model comparison;
- code snippets for R and Python that explain how to use the described methods.

On the other hand, **in this book, we do not focus on**

- any specific model. The techniques presented are model-agnostic and do not make any assumptions related to the model structure;
- data exploration. There are very good books on this topic by, for example, Grolemund and Wickham (2017) or Wes (2012), or the excellent classic by Tukey (1977);
- the process of model building. There are also very good books on this topic by, for instance, Venables and Ripley (2002), James et al. (2014), or Efron and Hastie (2016);
- any particular tools for model building. These are discussed, for instance, by Kuhn and Johnson (2013).

1.8 Acknowledgements

This book has been prepared by using the `bookdown` package (Xie, 2018), created thanks to the amazing work of Yihui Xie. A live version of this book is available at the GitHub repository https://github.com/pbiecek/ema. If you find any error, typo, or inaccuracy in the book, we will be grateful for your feedback at this website.

Figures and tables have been created mostly in the R language for statistical computing (R Core Team, 2018) with numerous libraries that support predictive modelling. Just to name a few packages frequently used in this book: `randomForest` (Liaw and Wiener, 2002), `ranger` (Wright and Ziegler, 2017), `rms` (Harrell Jr, 2018), `gbm` (Ridgeway, 2017), or `caret` (Kuhn, 2008). For statistical graphics, we have used the `ggplot2` package (Wickham, 2009). For model governance, we have used `archivist` (Biecek and Kosinski, 2017). Examples in Python were added thanks to the fantastic work of Hubert Baniecki and Wojciech Kretowicz, who develop and maintain the `dalex` library. Most of the presented examples concern models built in the `sklearn` library (Pedregosa et al., 2011). The `plotly` library (Plotly Technologies Inc., 2015) is used to visualize the results.

We would like to thank everyone who contributed with feedback, found typos, or ignited discussions while the book was being written, including GitHub contributors: Rees Morrison, Alicja Gosiewska, Kasia Pekala, Hubert Baniecki, Asia Henzel, Anna Kozak, Agile Bean, and Wojciech Kretowicz. We would like to acknowledge the anonymous reviewers, whose comments helped us to improve the contents of the book. We thank Jeff Webb, Riccardo De Bin, Patricia Martinkova, and Ziv Shkedy for their encouraging reviews. We are very grateful to John Kimmel from Chapman & Hall/CRC Press for his editorial assistance and patience.

Przemek's work on model interpretability started during research trips within the RENOIR (H2020 grant no. 691152) secondments to Nanyang Technological University (Singapour) and Davis University of California (USA). He would like to thank Prof. Janusz Holyst for the chance to take part in this project. Przemek would also like to thank Prof. Chris Drake for her hospitality. This book would have never been created without the perfect conditions that Przemek found at Chris's house in Woodland. Last but not least, Przemek would like to thank colleagues from the MI2DataLab and Samsung Research and Development Institute Poland for countless inspiring discussions related to Responsible Artificial Intelligence and Human Oriented Machine Learning.

Tomasz would like to thank colleagues from the Data Science Institute of Hasselt University and from the International Drug Development Institute (IDDI) for their support that allowed him finding the time to work on the book.

2

Model Development

2.1 Introduction

In general, we can distinguish between two approaches to statistical modelling: *explanatory* and *predictive* (Breiman, 2001b; Shmueli, 2010). In *explanatory modelling*, models are applied for inferential purposes, i.e., to test hypotheses resulting from some theoretical considerations related to the investigated phenomenon (for instance, related to an effect of a particular clinical factor on a probability of a disease). In *predictive modelling*, models are used for the purpose of predicting the value of a new or future observation (for instance, whether a person has got or will develop a disease). It is important to know what is the intended purpose of modelling because it has important consequences for the methods used in the model development process.

In this book, we focus on predictive modelling. Thus, we present mainly the methods relevant for predictive models. Nevertheless, we also show selected methods used in the case of explanatory models, in order to discuss, if relevant, substantive differences between the methods applied to the two approaches to modelling.

Predictive models are created for various purposes. For instance, a team of data scientists may spend months developing a single model that will be used for scoring risks of transactions in a large financial company. In that case, every aspect of the model is important, as the model will be used on a large scale and will have important long-term consequences for the company. Hence, the model-development process may be lengthy and tedious. On the other hand, if a small pizza-delivery chain wants to develop a simple model to roughly predict the demand for deliveries, the development process may be much shorter and less complicated. In that case, the model may be quickly updated or even discarded, without major consequences.

Irrespective of the goals of modelling, model-development process involves similar steps. In this chapter, we briefly discuss these steps.

2.2 Model-development process

Several approaches have been proposed to describe the process of model development. One of the most known general approaches is the Cross-industry Standard Process for Data Mining (CRISP-DM) (Chapman et al., 1999; Wikipedia, 2019). Methodologies specific for predictive models have been introduced also by Grolemund and Wickham (2017), Hall et al. (2019), and Biecek (2019).

The common goal of the approaches is to standardize the process. Standardization can help to plan resources needed to develop and maintain a model, and to make sure that no important steps are missed when developing the model.

CRISP-DM is a tool-agnostic procedure. It breaks the model-development process into six phases: *business understanding, data understanding, data preparation, modelling, evaluation,* and *deployment.* The phases can be iterated. Note that iterative phases are also considered by Grolemund and Wickham (2017) and Hall et al. (2019).

Figure 2.1 presents a variant of the iterative process, divided into five steps. Data collection and preparation is needed prior to any modelling. One cannot hope for building a model with good performance if the data are not of good quality. Once data have been collected, they have to be explored to understand their structure. Subsequently, a model can be selected and fitted to the data. The constructed model should be validated. The three steps: data understanding, model assembly, and model audit, are often iterated to arrive at a point when, for instance, a model with the best predictive performance is obtained. Once the "best" model has been obtained, it can be "delivered", i.e., implemented in practice after performing tests and developing the necessary documentation.

The Model-development Process (MDP), proposed by Biecek (2019), has been motivated by Rational Unified Process for Software Development (Kruchten, 1998; Jacobson et al., 1999; Boehm, 1988). MDP can be seen as an extension of the scheme presented in Figure 2.1. It recognizes that fact that consecutive iterations are not identical because the knowledge increases during the process and consecutive iterations are performed with different goals in mind. This is why MDP is presented in Figure 2.2 as an untangled version of Figure 2.1. The five phases, present in CRIPSP-DM, are shown in the rows. A single bar in each row represents a number of resources (for instance, a week-worth workload) that can be devoted to the project at a specific time-point (indicated on the horizontal axis). For a particular phase, resources can be used in different amounts depending on the current stage of the process, as indicated by the height of the bars. The stages are indicated at the top of the diagram in Figure 2.2: *problem formulation, crisp modelling, fine tuning,* and *maintenance and decommissioning.* Problem formulation aims at defining the needs for the model,

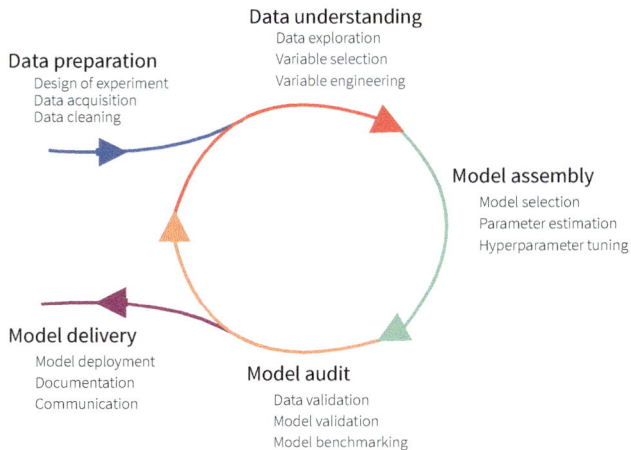

Data understanding
Data exploration
Variable selection
Variable engineering

Data preparation
Design of experiment
Data acquisition
Data cleaning

Model assembly
Model selection
Parameter estimation
Hyperparameter tuning

Model delivery
Model deployment
Documentation
Communication

Model audit
Data validation
Model validation
Model benchmarking

FIGURE 2.1 The lifecycle of a predictive model.

defining datasets that will be used for training and validation, and deciding which performance measures will be used for the evaluation of the performance of the final model. Crisp modelling focuses on the creation of first versions of the model that may provide an idea about, for instance, how complex may the model have to be to yield the desired solution? Fine-tuning focuses on improving the initial version(s) of the model and selecting the best one according to the pre-defined metrics. Finally, maintenance and decommissioning aims at monitoring the performance of the model after its implementation. Note that, unlike in CRISP-DM, the diagram in Figure 2.2 indicates that the process may start with some resources being spent not on the data-preparation phase, but on the model-audit one. This is because, at the problem formulation stage, we may have to spend some time on defining the goals and model-performance metrics (that will be used in model benchmarking) before any attempt to collect the data.

Figure 2.2 also indicates that there may be several iterations of the different phases within each stage, as indicated at the bottom of the diagram. For instance, in the crisp-modelling stage, several versions of a model may be prepared in subsequent iterations.

Methods presented in this book can be used to better understand the data and the application domain (*exploration*), obtain insight into model-based predictions (*model explanation*), and evaluate a model's performance (*model*

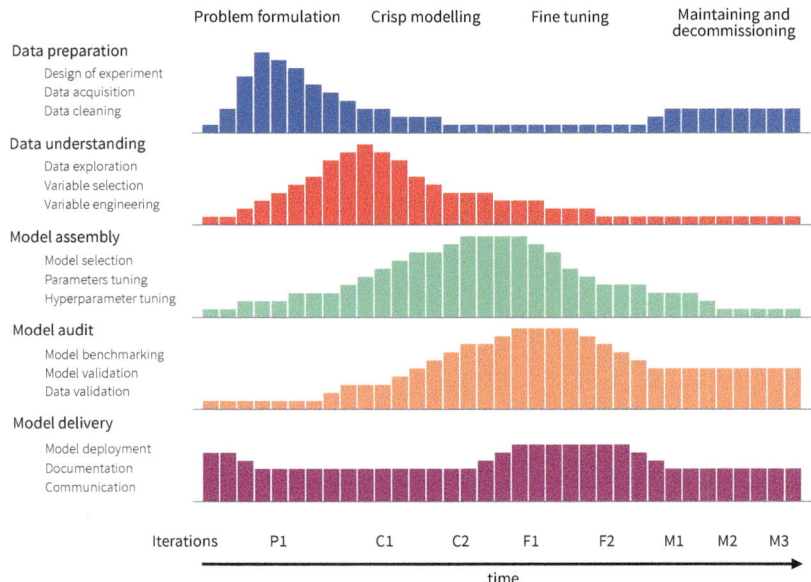

FIGURE 2.2 Overview of the model-development process. The process is split into five different phases (rows) and four stages (indicated at the top of the diagram). Horizontal axis presents the time from the problem formulation to putting the model into practice (decommissioning). For a particular phase, resources can be used in different amounts depending on the current stage of the process, as indicated by the height of the bars. There may be several iterations of different phases within each stage, as indicated at the bottom of the diagram.

examination). Thus, referring to MDP in Figure 2.2, the methods are suitable for data understanding, model assembly, and model audit phases.

In the remainder of the chapter, we provide a brief overview of the notation that will be used in the book, and the methods commonly used for data exploration, model fitting, and model validation.

2.3   Notation

Methods described in this book have been developed by different authors, who used different mathematical notations. We have made an attempt at keeping the mathematical notation consistent throughout the entire book. In some cases this may result in formulas with a fairly complex system of indices.

In this section, we provide a general overview of the notation we use. Whenever necessary, parts of the notation will be explained again in subsequent chapters.

We use capital letters like X or Y to denote (scalar) random variables. Observed values of these variables are denoted by lower case letters like x or y. Vectors and matrices are distinguished by underlining the letter. Thus, $\underline{X}$ and $\underline{x}$ denote matrix X and (column) vector x, respectively. Note, however, that in some situations $\underline{X}$ may indicate a vector of (scalar) random variables. We explicitly mention this when needed. Transposition is indicated by the prime, i.e., $\underline{x}'$ is the row vector resulting from transposition of a column vector $\underline{x}$.

We use notation $E(Y)$ and $Var(Y)$ to denote the expected (mean) value and variance of random variable Y. If needed, we use a subscript to indicate the distribution used to compute the parameters. Thus, for instance, we use

$$E_{Y|X=x}(Y) = E_{Y|x}(Y) = E_Y(Y|X = x)$$

to indicate the conditional mean of Y given that random variable X assumes the value of x.

We assume that the data available for modelling consist of n observations/instances. For the i-th observation, we have got an observed value of y_i of a dependent (random) variable Y. We assume that Y is a scalar, i.e., a single number. In case of dependent categorical variable, we usually consider Y to be a binary indicator of observing a particular category.

Additionally, each observation from a dataset is described by p explanatory variables. We refer to the (column) vector of the explanatory variables, describing the i-th observation, by $\underline{x}_i$. We can thus consider observations as points in a p-dimensional space $\mathcal{X} \equiv \mathcal{R}^p$, with $\underline{x}_i \in \mathcal{X}$. We often collect all explanatory-variable data in the $n \times p$ matrix $\underline{X}$ that contains, in the i-th row, vector $\underline{x}_i'$.

When introducing some of the model-exploration methods, we often consider "an observation of interest", for which the vector of explanatory variables is denoted by x_*. As the observation may not necessarily belong to the analyzed dataset, we use the asterisk in the subscript. Clearly, $\underline{x}_* \in \mathcal{X}$.

We refer to the j-th coordinate of vector $\underline{x}$ by using j in superscript. Thus, $\underline{x}_i = (x_i^1, \ldots, x_i^p)'$, where x_i^j denotes the j-th coordinate of vector $\underline{x}_i$ for the i-th observation from the analyzed dataset. If a power (for instance, a square) of x_i^j is needed, it will be denoted by using parentheses, i.e., $\left(x_i^j\right)^2$.

If $\mathcal{J}$ denotes a subset of indices, then $\underline{x}^{\mathcal{J}}$ denotes the vector formed by the coordinates of $\underline{x}$ corresponding to the indices included in $\mathcal{J}$. We use $\underline{x}^{-j}$ to refer to a vector that results from removing the j-th coordinate from vector

$\underline{x}$. By $\underline{x}^{j|=z}$, we denote a vector in which all coordinates are equal to their values in $\underline{x}$, except of the j-th coordinate, whose value is set equal to z. Thus, $\underline{x}^{j|=z} = (x^1, \ldots, x^{j-1}, z, x^{j+1}, \ldots, x^p)'$.

Notation $\underline{X}^{*j}$ is used to denote a matrix with the values as in $\underline{X}$ except of the j-th column, for which elements are permuted.

In this book, a model is a function $f : \mathcal{X} \rightarrow \mathcal{R}$ that transforms a point from $\mathcal{X}$ into a real number. In most cases, the presented methods can be used directly for multivariate dependent variables; however, we use examples with univariate responses to simplify the notation.

Typically, during the model development, we create many competing models. Formally, we shall index models to refer to a specific version fitted to a dataset. However, for the sake of simplicity, we will omit the index when it is not important. For the same reason we ignore in the notation the fact that, in practice, we never know true model coefficients and use the estimated values.

We use the term "model residual" to indicate the difference between the observed value of the dependent variable Y for the i-th observation from a particular dataset and the model's prediction for the observation:

$$r_i = y_i - f(x_i) = y_i - \hat{y}_i, \tag{2.1}$$

where $\hat{y}_i$ denotes the predicted (or fitted) value of y_i. More information about residuals is provided in Chapter 19.

2.4   Data understanding

As indicated in Figures 2.1 and 2.2, before starting construction of any models, we have got to understand the data. Toward this aim, tools for data exploration, such as visualization techniques, tabular summaries, and statistical methods can be used. The choice of the tools depends on the character of variables included in a dataset.

The most known introduction to data exploration is a famous book by Tukey (1977). It introduces the (now classical) tools like, for example, box-and-whisker plots or stem-and-leaf plots. Good overviews of techniques for data exploration can also be found in books by Nolan and Lang (2015) and Wickham and Grolemund (2017).

In this book, we rely on five visualization techniques for data exploration, schematically presented in Figure 2.3. Two of them (histogram and empirical cumulative-distribution (ECD) plot) are used to summarize the distribution of a single random (explanatory or dependent) variable; the remaining three (mosaic plot, box plot, and scatter plot) are used to explore the relationship between pairs of variables. Note that a histogram can be used to explore the distribution of a continuous or a categorical variable, while ECD and box plots are suitable for continuous variables. A mosaic plot is useful for exploring the relationship between two categorical variables, while a scatter plot can be applied for two continuous variables. It is worth noting that box plots can also be used for evaluating a relation between a categorical variable and a continuous one, as illustrated in Figure 2.3.

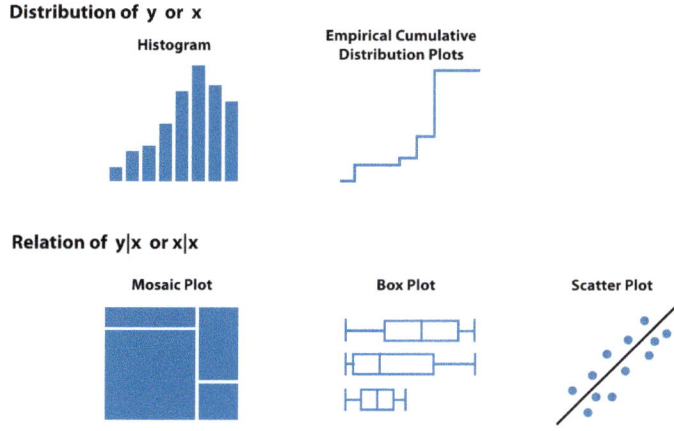

FIGURE 2.3 Selected methods for visual data exploration applied in this book.

Exploration of data for the dependent variable usually focuses on the question related to the distribution of the variable. For instance, for a continuous variable, questions like approximate normality or symmetry of the distribution are most often of interest, because of the availability of many powerful methods and models that use the normality assumption. In the case of asymmetry (skewness), a possibility of a transformation that could make the distribution approximately symmetric or normal is usually investigated. For a categorical dependent variable, an important question is whether the proportion of observations in different categories is balanced or not. This is because, for instance, some methods related to the classification problem do not work well with if there is a substantial imbalance between the categories.

Exploration of data for explanatory variables may also include investigation of their distribution. This is because the results may reveal, for instance, that there is little variability in the observed values of a variable. As a consequence, the variable might be deemed not interesting from a model-construction point of view. Usually, however, the exploration focuses on the relationship between explanatory variables themselves on one hand, and their relationship with the dependent variable on the other hand. The results may have important consequences for model construction. For instance, if an explanatory variable does not appear to be related to the dependent variable, it may be dropped from a model (*variable selection/filtering*). The exploration results may also suggest, for instance, a need for a transformation of an explanatory variable to make its relationship with the dependent variable linear (*variable engineering*). Detection of pairs of strongly-correlated explanatory variables is also of interest, as it may help in resolving issues with, for instance, instability of optimization algorithms used for fitting of a model.

2.5 Model assembly (fitting)

In statistical modelling, we are interested in the distribution of a dependent variable Y given $\underline{x}$, the vector of values of explanatory variables. In the ideal world, we would like to know the entire conditional distribution. In practical applications, however, we usually do not evaluate the entire distribution, but just some of its characteristics, like the expected (mean) value, a quantile, or variance. Without loss of generality we will assume that we model the conditional expected value of Y, i.e., $E_{Y|\underline{x}}(Y)$.

Assume that we have got model $f()$, for which $f(\underline{x})$ is an approximation of $E_{Y|\underline{x}}(Y)$, i.e., $E_{Y|\underline{x}}(Y) \approx f(\underline{x})$. Note that, in our book, we do not assume that it is a "good" model, nor that the approximation is precise. We simply assume that we have got a model that is used to estimate the conditional expected value and to form predictions of the values of the dependent variable. Our interest lies in the evaluation of the quality of the predictions. If the model offers a "good" approximation of the conditional expected value, it should be reflected in its satisfactory predictive performance.

Usually, when building a model, the available data are split into two parts. One part, often called a "training set" or "learning data", is used for estimation of the model coefficients. The other part, called a "testing set" or "validation data", is used for model validation. The splitting may be done repeatedly, as in k-fold cross-validation. We leave the topic of model validation for Chapter 15.

The process of estimation of model coefficients based on the training data, i.e., "fitting" of the model, differs for different models. In most cases, however, it can be seen as an optimization problem. Let Θ be the space of all possible values of model coefficients. Model fitting (or training) is a procedure of selecting a value $\underline{\hat{\theta}} \in \Theta$ that minimizes some loss function $L()$:

$$\underline{\hat{\theta}} = \arg \min_{\underline{\theta} \in \Theta} L\{\underline{y}, f(\underline{\theta}; \underline{X})\}, \tag{2.2}$$

where $\underline{y}$ is the vector of observed values of the dependent variable and $f(\underline{\theta}; \underline{X})$

is the corresponding vector of the model's predictions computed for model coefficients $\underline{\theta}$ and matrix $\underline{X}$ of values of explanatory variables for the observations from the training dataset. Denote the estimated form of the model by $f(\hat{\underline{\theta}}; \underline{X})$.

Consider predction of a new observation for which the vector of explanatory variables assumes the value of $\underline{x}_*$, i.e., $f(\hat{\underline{\theta}}; \underline{x}_*)$. Assume that $E_{Y|\underline{x}_*}(Y) = f(\underline{\theta}; \underline{x}_*)$. It can be shown (Hastie et al., 2009; Shmueli, 2010) that the expected squared-error of prediction can be expressed as follows:

$$E_{(Y,\hat{\underline{\theta}})|\underline{x}_*}\{Y - f(\hat{\underline{\theta}}; \underline{x}_*)\}^2 = E_{Y|\underline{x}_*}\{Y - f(\underline{\theta}; \underline{x}_*)\}^2 +$$
$$[f(\underline{\theta}; \underline{x}_*) - E_{\hat{\underline{\theta}}|\underline{x}_*}\{f(\hat{\underline{\theta}}; \underline{x}_*)\}]^2 +$$
$$E_{\hat{\underline{\theta}}|\underline{x}_*}[f(\hat{\underline{\theta}}; \underline{x}_*) - E_{\hat{\underline{\theta}}|\underline{x}_*}\{f(\hat{\underline{\theta}}; \underline{x}_*)\}]^2$$
$$= Var_{Y|\underline{x}_*}(Y) + Bias^2 + Var_{\hat{\underline{\theta}}|\underline{x}_*}\{\hat{f}(\underline{x}_*)\}. \quad (2.3)$$

The first term on the right-hand-side of equation (2.3) is the variability of Y around its conditional expected value $f(\underline{\theta}; \underline{x}_*)$. In general, it cannot be reduced. The second term is the squared difference between the expected value and its estimate, i.e., the squared bias. Bias results from misspecifying the model by, for instance, using a more parsimonious or a simpler model. The third term is the variance of the estimate, due to the fact that we use training data to estimate the model.

The decomposition presented in (2.3) underlines an important difference between explanatory and predictive modelling. In the explanatory modelling, the goal is to minimize the bias, as we are interested in obtaining the most accurate representation of the investigated phenomenon and the related theory. In the predictive modelling, the focus is on minimization of the sum of the (squared) bias and the estimation variance, because we are interested in minimization of the prediction error. Thus, sometimes we can accept a certain amount of bias, if it leads to a substantial gain in precision of estimation and, consequently, in a smaller prediction error (Shmueli, 2010).

It follows that the choice of the loss function $L()$ in equation (2.2) may differ for explanatory and predictive modelling. For the former, it is common to assume some family of probability distributions for the conditional distribution of Y given $\underline{x}$. In such case, the loss function $L()$ may be defined as the negative logarithm of the likelihood function, where the likelihood is the probability of observing $\underline{y}$, given $\underline{X}$, treated as a function of $\underline{\theta}$. The resulting estimate of $\underline{\theta}$ is usually denoted by $\hat{\underline{\theta}}$.

In predictive modelling, it is common to add term $\lambda(\underline{\theta})$ to the loss function that "penalizes" for the use of more complex models:

$$\tilde{\underline{\theta}} = \arg\min_{\underline{\theta} \in \Theta} \left[L\{\underline{y}, f(\underline{\theta}; \underline{X})\} + \lambda(\underline{\theta}) \right]. \quad (2.4)$$

For example, in linear regression we assume that the observed vector $\underline{y}$ follows a multivariate normal distribution:

$$\underline{y} \sim \mathcal{N}(\underline{X}'\underline{\beta}, \sigma^2 \underline{I}_n),$$

where $\underline{\theta}' = (\underline{\beta}', \sigma^2)$ and $\underline{I}_n$ denotes the $n \times n$ identity matrix. In this case, equation (2.4) becomes

$$\underline{\tilde{\theta}} = \arg\min_{\underline{\theta} \in \Theta} \left\{ \frac{1}{n} ||\underline{y} - \underline{X}'\underline{\beta}||_2 + \lambda(\underline{\beta}) \right\} = \arg\min_{\underline{\theta} \in \Theta} \left\{ \frac{1}{n} \sum_{i=1}^{n} (y_i - \underline{x}_i'\underline{\beta})^2 + \lambda(\underline{\beta}) \right\}.$$
$$(2.5)$$

For the classcal linear regression, the penalty term $\lambda(\underline{\beta})$ is equal to 0. In that case, the optimal parameters $\underline{\hat{\beta}}$ and $\hat{\sigma}^2$, obtained from (2.2), can be expressed in a closed form:

$$\underline{\hat{\beta}} = (\underline{X}'\underline{X})^{-1}\underline{X}'\underline{y},$$

$$\hat{\sigma}^2 = \frac{1}{n}||\underline{y} - \underline{X}'\underline{\hat{\beta}}||_2 = \frac{1}{n}\sum_{i=1}^{n}(y_i - \underline{x}_i'\underline{\hat{\beta}})^2 = \frac{1}{n}\sum_{i=1}^{n}(y_i - \hat{y}_i)^2.$$

On the other hand, in ridge regression, the penalty function is defined as follows:

$$\lambda(\underline{\beta}) = \lambda \cdot ||\underline{\beta}||_2 = \lambda \sum_{k=1}^{p}(\beta^k)^2. \qquad (2.6)$$

In that case, the optimal parameters $\underline{\tilde{\beta}}$ and $\tilde{\sigma}^2$, obtained from equation (2.4), can also be expressed in a closed form:

$$\underline{\tilde{\beta}} = (\underline{X}'\underline{X} + \lambda\underline{I}_n)^{-1}\underline{X}'\underline{y},$$

$$\tilde{\sigma}^2 = \frac{1}{n}||\underline{y} - \underline{X}'\underline{\tilde{\beta}}||_2.$$

Note that ridge regression leads to non-zero squared-bias in equation (2.3), but at the benefit of a reduced estimation variance (Hastie et al., 2009).

Another possible form of penalty, used in the Least Absolute Shrinkage and Selection Operator (LASSO) regression, is given by

$$\lambda(\underline{\beta}) = \lambda \cdot ||\underline{\beta}||_1 = \lambda \sum_{k=1}^{p}|\beta^k|. \qquad (2.7)$$

In that case, $\tilde{\underline{\beta}}$ and $\tilde{\sigma}^2$ have to be obtained by using a numerical optimization procedure.

For a binary dependent variable, i.e., a classification problem, the natural choice for the distribution of Y is the Bernoulli distribution. The resulting loss function, based on the logarithm of the Bernoulli likelihood, is

$$L(\underline{y}, \underline{p}) = -\frac{1}{n} \sum_{i=1}^{n} \{ y_i \ln p_i + (1 - y_i) \ln (1 - p_i) \}, \qquad (2.8)$$

where y_i is equal to 0 or 1 in case of "no response" and "response" (or "failure" and "success"), and p_i is the probability of y_i being equal to 1. Function (2.8) is often called "log-loss" or "binary cross-entropy" in machine-learning literature.

A popular model for binary data is logistic regression, for which

$$\ln \frac{p_i}{1 - p_i} = \underline{x}'_i \underline{\beta}.$$

In that case, the loss function in equation (2.8) becomes equal to

$$L\{\underline{y}, f(\underline{\beta}, \underline{X})\} = -\frac{1}{n} \sum_{i=1}^{n} [y_i \underline{x}'_i \underline{\beta} - \ln\{1 + \exp(\underline{x}'_i \underline{\beta})\}]. \qquad (2.9)$$

Optimal values of parameters $\hat{\underline{\beta}}$, resulting from equation (2.2), have to be found by numerical optimization algorithms.

Of course, one can combine the loss functions in equations (2.8) and (2.9) with penalties (2.6) or (2.7) .

For a categorical dependent variable, i.e., a multilabel classification problem, the natural choice for the distribution of Y is the multinomial distribution. The resulting loss function, in case of K categories, is given by

$$L(\underline{Y}, \underline{P}) = -\frac{1}{n} \sum_{i=1}^{n} \sum_{k=1}^{K} y_{ik} \ln p_{ik}, \qquad (2.10)$$

where $y_{ik} = 1$ if the k-th category was noted for the i-th observation and 0 otherwise, and p_{ik} is the probability of y_{ik} being equal to 1. Function (2.10) is often called "categorical cross-entropy" in machine-learning literature. Also in this case, optimal parameters $\hat{\underline{\beta}}$, resulting from equation (2.2), have to be found by numerical optimization algorithms.

2.6   Model audit

As indicated in Figure 2.2, the modelling process starts with some crisp early versions that are fine-tuned in consecutive iterations. To arrive at a final model, we usually have got to evaluate (audit) numerous candidate models that. In this book, we introduce techniques that allow:

- decomposing a model's predictions into components that can be attributed to particular explanatory variables (Chapters 6–9).
- conducting sensitivity analysis for a model's predictions (Chapter 10–12).
- summarizing the predictive performance of a model (Chapter 15). In particular, the presented measures are usually used to trace the progress in model development.
- assessing the importance of an explanatory variable (Chapter 16). The techniques can be helpful in reducing the set of explanatory variables to be included in a model in the fine-tuning stage.
- evaluating the effect of an explanatory variable on a model's predictions (Chapters 17–18).
- detailed examination of both overall and instance-specific model performance (Chapter 19). These are residual-diagnostic tools that can help in identifying potential causes that may lead to issues with model performance.

All those techniques can be used to evaluate the current version of a model and to get suggestions for possible improvements. The improvements may be developed and evaluated in the next crisp-modelling or fine-tuning phase.

3

Do-it-yourself

Most of the methods presented in this book are available in both R and Python and can be used in a uniform way. But each of these languages has also many other tools for Explanatory Model Analysis.

In this book, we introduce various methods for instance-level and dataset-level exploration and explanation of predictive models. In each chapter, there is a section with code snippets for R and Python that shows how to use a particular method.

3.1 Do-it-yourself with R

In this section, we provide a short description of the steps that are needed to set-up the R environment with the required libraries.

3.1.1 What to install?

Obviously, the R software (R Core Team, 2018) is needed. It is always a good idea to use the newest version. At least R in version 3.6 is recommended. It can be downloaded from the CRAN website https://cran.r-project.org/.

A good editor makes working with R much easier. There are plenty of choices, but, especially for beginners, consider the RStudio editor, an open-source and enterprise-ready tool for R. It can be downloaded from https://www.rstudio.com/.

Once R and the editor are available, the required packages should be installed.

The most important one is the DALEX package in version 1.0 or newer. It is the entry point to solutions introduced in this book. The package can be installed by executing the following command from the R command line:

```
install.packages("DALEX")
```

Installation of DALEX will automatically take care about installation of other requirements (packages required by it), like the ggplot2 package for data visualization, or `ingredients` and `iBreakDown` with specific methods for model exploration.

3.1.2   How to work with `DALEX`?

To conduct model exploration with `DALEX`, first, a model has to be created. Then the model has got to be prepared for exploration.

There are many packages in R that can be used to construct a model. Some packages are algorithm-specific, like `randomForest` for random forest classification and regression models (Liaw and Wiener, 2002), `gbm` for generalized boosted regression models (Ridgeway, 2017), `rms` with extensions for generalized linear models (Harrell Jr, 2018), and many others. There are also packages that can be used for constructing models with different algorithms; these include the `h2o` package (LeDell et al., 2019), `caret` (Kuhn, 2008) and its successor `parsnip` (Kuhn and Vaughan, 2019), a very powerful and extensible framework `mlr` (Bischl et al., 2016), or `keras` that is a wrapper to Python library with the same name (Allaire and Chollet, 2019).

While it is great to have such a large choice of tools for constructing models, the disadvantage is that different packages have different interfaces and different arguments. Moreover, model-objects created with different packages may have different internal structures. The main goal of the `DALEX` package is to create a level of abstraction around a model that makes it easier to explore and explain the model. Figure 3.1 illustrates the contents of the package. In particular, function `DALEX::explain` is THE function for model wrapping. There is only one argument that is required by the function; it is `model`, which is used to specify the model-object with the fitted form of the model. However, the function allows additional arguments that extend its functionalities. They are discussed in Section 4.2.6.

3.1.3   How to work with `archivist`?

As we will focus on the exploration of predictive models, we prefer not to waste space nor time on replication of the code necessary for model development. This is where the `archivist` packages help.

The `archivist` package (Biecek and Kosinski, 2017) is designed to store, share, and manage R objects. We will use it to easily access R objects for pre-constructed models and pre-calculated explainers. To install the package, the following command should be executed in the R command line:

```
install.packages("archivist")
```

Once the package has been installed, function `aread()` can be used to retrieve R objects from any remote repository. For this book, we use a GitHub repository `models` hosted at https://github.com/pbiecek/models. For instance, to

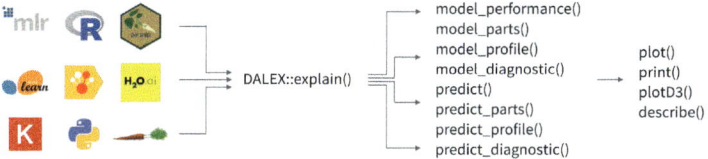

FIGURE 3.1 The `DALEX` package creates a layer of abstraction around models, allowing you to work with different models in a uniform way. The key function is the `explain()` function, which wraps any model into a uniform interface. Then other functions from the `DALEX` package can be applied to the resulting object to explore the model.

download a model with the md5 hash `ceb40`, the following command has to be executed:

```
archivist::aread("pbiecek/models/ceb40")
```

Since the md5 hash `ceb40` uniquely defines the model, referring to the repository object results in using exactly the same model and the same explanations. Thus, in the subsequent chapters, pre-constructed models will be accessed with `archivist` hooks. In the following sections, we will also use `archivist` hooks when referring to datasets.

3.2 Do-it-yourself with Python

In this section, we provide a short description of steps that are needed to set-up the Python environment with the required libraries.

3.2.1 What to install?

The Python interpreter (van Rossum and Drake, 2009) is needed. It is always a good idea to use the newest version. Python in version 3.6 is the minimum recommendation. It can be downloaded from the Python website https://python.org/. A popular environment for a simple Python installation and configuration is Anaconda, which can be downloaded from website https://www.anaconda.com/.

There are many editors available for Python that allow editing the code in a convenient way. In the data science community a very popular solution is Jupyter Notebook. It is a web application that allows creating and sharing documents that contain live code, visualizations, and descriptions. Jupyter Notebook can be installed from the website https://jupyter.org/.

Once Python and the editor are available, the required libraries should be installed. The most important one is the `dalex` library, currently in version `0.2.0`. The library can be installed with `pip` by executing the following instruction from the command line:

```
pip install dalex
```

Installation of `dalex` will automatically take care of other required libraries.

3.2.2 How to work with `dalex`?

There are many libraries in Python that can be used to construct a predictive model. Among the most popular ones are algorithm-specific libraries like `catboost` (Dorogush et al., 2018), `xgboost` (Chen and Guestrin, 2016), and `keras` (Gulli and Pal, 2017), or libraries with multiple ML algorithms like `scikit-learn` (Pedregosa et al., 2011).

While it is great to have such a large choice of tools for constructing models, the disadvantage is that different libraries have different interfaces and different arguments. Moreover, model-objects created with different library may have different internal structures. The main goal of the `dalex` library is to create a level of abstraction around a model that makes it easier to explore and explain the model.

Constructor `Explainer()` is THE method for model wrapping. There is only one argument that is required by the function; it is `model`, which is used to specify the model-object with the fitted form of the model. However, the function also takes additional arguments that extend its functionalities. They are discussed in Section 4.3.6. If these additional arguments are not provided by the user, the `dalex` library will try to extract them from the model. It is a good idea to specify them directly to avoid surprises.

As soon as the model is wrapped by using the `Explainer()` function, all further functionalities can be performed on the resulting object. They will be presented in subsequent chapters in subsections *Code snippets for Python*.

3.2.3 Code snippets for Python

A detailed description of model exploration will be presented in the next chapters. In general, however, the way of working with the `dalex` library can be described in the following steps:

1. Import the `dalex` library.

```
import dalex as dx
```

2. Create an **Explainer** object. This serves as a wrapper around the model.

```python
exp = dx.Explainer(model, X, y)
```

3. Calculate predictions for the model.

```python
exp.predict(henry)
```

4. Calculate specific explanations.

```python
obs_bd = exp.predict_parts(obs, type='break_down')
```

5. Print calculated explanations.

```python
obs_bd.result
```

6. Plot calculated explanations.

```python
obs_bd.plot()
```

4

Datasets and Models

We will illustrate the methods presented in this book by using three datasets related to:

- predicting probability of survival for passengers of the *RMS Titanic*;
- predicting prices of *apartments in Warsaw*;
- predicting the value of the football players based on the *FIFA* dataset.

The first dataset will be used to illustrate the application of the techniques in the case of a predictive (classification) model for a binary dependent variable. It is mainly used in the examples presented in the second part of the book. The second dataset will be used to illustrate the exploration of prediction models for a continuous dependent variable. It is mainly used in the examples in the third part of this book. The third dataset will be introduced in Chapter 21 and will be used to illustrate the use of all of the techniques introduced in the book.

In this chapter, we provide a short description of the first two datasets, together with results of exploratory analyses. We also introduce models that will be used for illustration purposes in subsequent chapters.

4.1 Sinking of the RMS Titanic

The sinking of the RMS Titanic is one of the deadliest maritime disasters in history (during peacetime). Over 1500 people died as a consequence of a collision with an iceberg. Projects like *Encyclopedia Titanica* (https://www.encyclopedia-titanica.org/) are a source of rich and precise data about Titanic's passengers. The `stablelearner` package in R includes a data frame with information about passengers' characteristics. The dataset, after some data cleaning and variable transformations, is also available in the DALEX package for R and in the `dalex` library for Python. In particular, the `titanic` data frame contains 2207 observations (for 1317 passengers and 890 crew members) and nine variables:

- *gender*, person's (passenger's or crew member's) gender, a factor (categorical variable) with two levels (categories): "male" (78%) and "female" (22%);

- *age*, person's age in years, a numerical variable; the age is given in (integer) years, in the range of 0–74 years;
- *class*, the class in which the passenger travelled, or the duty class of a crew member; a factor with seven levels: "1st" (14.7%), "2nd" (12.9%), "3rd" (32.1%), "deck crew" (3%), "engineering crew" (14.7%), "restaurant staff" (3.1%), and "victualling crew" (19.5%);
- *embarked*, the harbor in which the person embarked on the ship, a factor with four levels: "Belfast" (8.9%), "Cherbourg" (12.3%), "Queenstown" (5.6%), and "Southampton" (73.2%);
- *country*, person's home country, a factor with 48 levels; the most common levels are "England" (51%), "United States" (12%), "Ireland" (6.2%), and "Sweden" (4.8%);
- *fare*, the price of the ticket (only available for passengers; 0 for crew members), a numerical variable in the range of 0–512;
- *sibsp*, the number of siblings/spouses aboard the ship, a numerical variable in the range of 0–8;
- *parch*, the number of parents/children aboard the ship, a numerical variable in the range of 0–9;
- *survived*, a factor with two levels: "yes" (67.8%) and "no" (32.2%) indicating whether the person survived or not.

The first six rows of this dataset are presented in the table below.

gender	age	class	embarked	fare	sibsp	parch	survived
male	42	3rd	Southampton	7.11	0	0	no
male	13	3rd	Southampton	20.05	0	2	no
male	16	3rd	Southampton	20.05	1	1	no
female	39	3rd	Southampton	20.05	1	1	yes
female	16	3rd	Southampton	7.13	0	0	yes
male	25	3rd	Southampton	7.13	0	0	yes

Models considered for this dataset will use *survived* as the (binary) dependent variable.

4.1.1   Data exploration

As discussed in Chapter 2, it is always advisable to explore data before modelling. However, as this book is focused on model exploration, we will limit the data exploration part.

Before exploring the data, we first conduct some pre-processing. In particular, the value of variables *age, country, sibsp, parch*, and *fare* is missing for a limited number of observations (2, 81, 10, 10, and 26, respectively). Analyzing data with missing values is a topic on its own (Schafer, 1997; Little and Rubin, 2002; Molenberghs and Kenward, 2007). An often-used approach is to impute the missing values. Toward this end, multiple imputations should be considered (Schafer, 1997; Molenberghs and Kenward, 2007; van Buuren, 2012). However, given the limited number of missing values and the intended illustrative use of the dataset, we will limit ourselves to, admittedly inferior, single imputation. In particular, we replace the missing *age* values by the mean of the observed ones, i.e., 30. Missing *country* is encoded by "X". For *sibsp* and *parch*, we replace the missing values by the most frequently observed value, i.e., 0. Finally, for *fare*, we use the mean fare for a given *class*, i.e., 0 pounds for crew, 89 pounds for the first, 22 pounds for the second, and 13 pounds for the third class.

After imputing the missing values, we investigate the association between survival status and other variables. Most variables in the Titanic dataset are categorical, except of *age* and *fare*. Figure 4.1 shows histograms for the latter two variables. In order to keep the exploration uniform, we transform the two variables into categorical ones. In particular, *age* is discretized into five categories by using cutoffs equal to 5, 10, 20, and 30, while *fare* is discretized by applying cutoffs equal to 1, 10, 25, and 50.

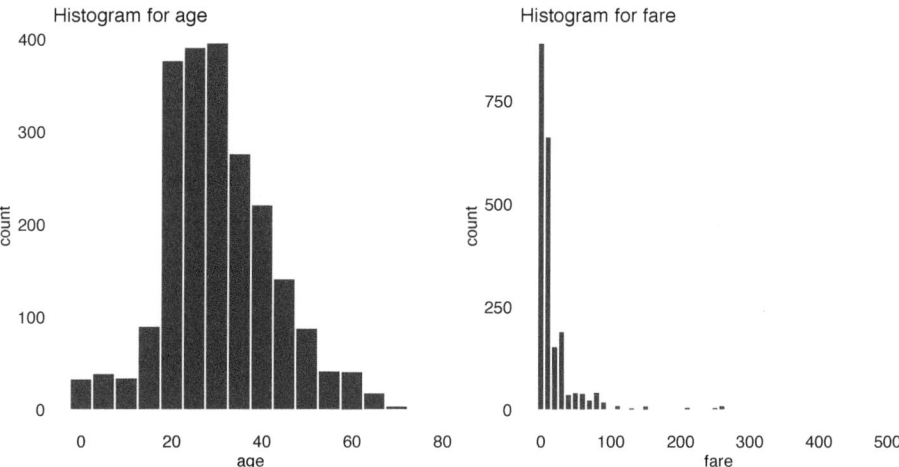

FIGURE 4.1 Histograms for variables *age* and *fare* from the Titanic data.

Figures 4.2–4.5 present graphically, with the help of mosaic plots, the proportion of non- and survivors for different levels of other variables. The width of the bars (on the x-axis) reflects the marginal distribution (proportions) of the observed levels of the variable. On the other hand, the height of the bars (on the y-axis) provides information about the proportion of non- and survivors. The graphs for *age* and *fare* were constructed by using the categorized versions of the variables.

Figure 4.2 indicates that the proportion of survivors was larger for females and children below 5 years of age. This is most likely the result of the "women and children first" principle that is often evoked in situations that require the evacuation of persons whose life is in danger.

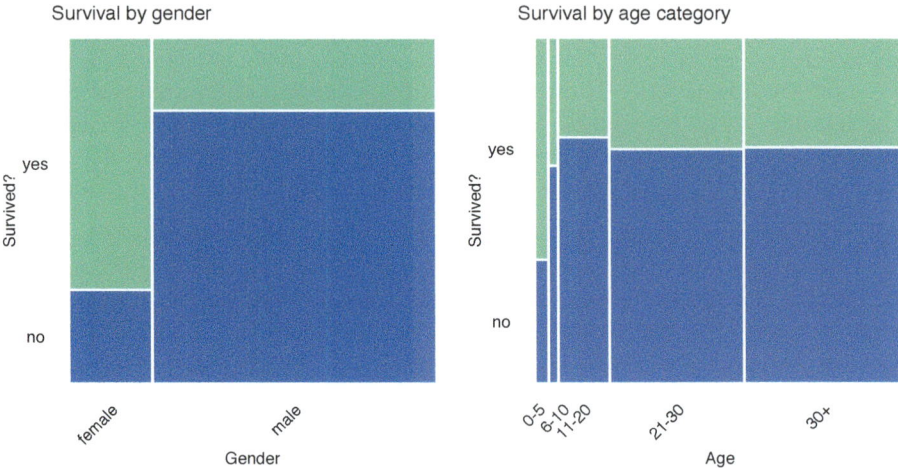

FIGURE 4.2 Survival according to gender and age category in the Titanic data.

The principle can, perhaps, partially explain the trend seen in Figure 4.3, i.e., a higher proportion of survivors among those with 1-2 parents/children and 1-2 siblings/spouses aboard.

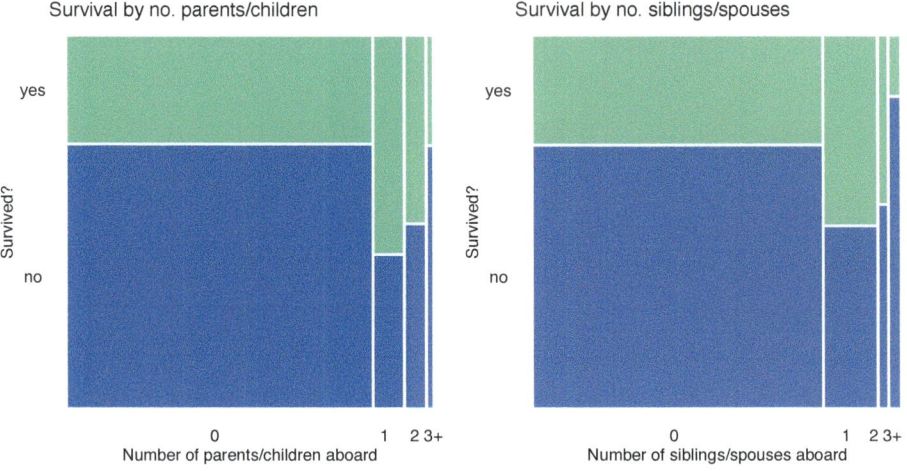

FIGURE 4.3 Survival according to the number of parents/children and siblings/spouses in the Titanic data.

Figure 4.4 indicates that passengers travelling in the first and second class had a higher chance of survival, perhaps due to the proximity of the location of their cabins to the deck. Interestingly, the proportion of survivors among the deck crew was similar to the proportion of the first-class passengers. The figure also shows that the proportion of survivors increased with the fare, which is consistent with the fact that the proportion was higher for passengers travelling in the first and second class.

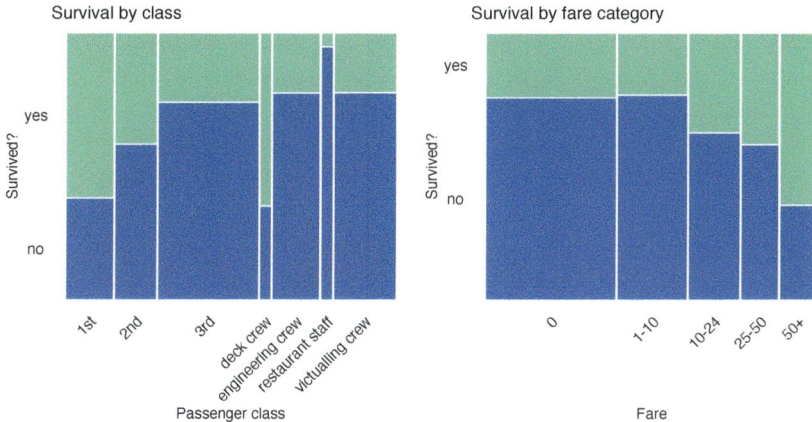

FIGURE 4.4 Survival according to travel-class and ticket-fare in the Titanic data.

Finally, Figure 4.5 does not suggest any noteworthy trends.

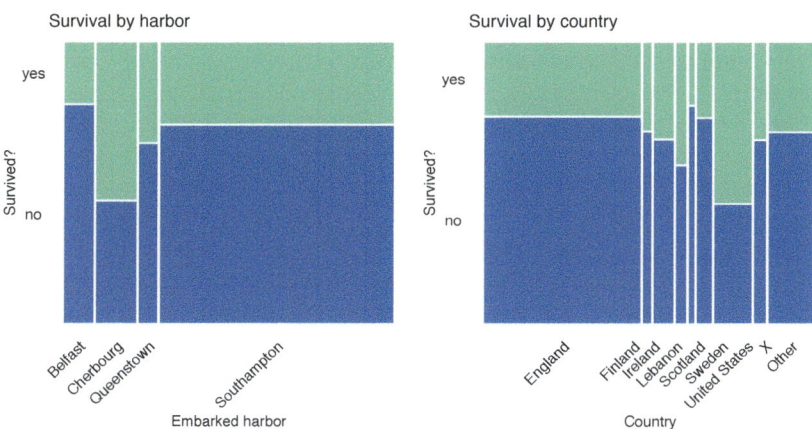

FIGURE 4.5 Survival according to the embarked harbour and country in the Titanic data.

4.2   Models for RMS Titanic, snippets for R

4.2.1   Logistic regression model

The dependent variable of interest, *survived*, is binary. Thus, a natural choice is to start the predictive modelling with a logistic regression model. As there is no reason to expect a linear relationship between age and odds of survival, we use linear tail-restricted cubic splines, available in the `rcs()` function of the `rms` package (Harrell Jr, 2018), to model the effect of age. We also do not expect linear relation for the *fare* variable, but because of its skewness (see Figure 4.1), we do not use splines for this variable. The results of the model are stored in model-object `titanic_lmr`, which will be used in subsequent chapters.

```
library("rms")
titanic_lmr <- lrm(survived == "yes" ~ gender + rcs(age) + class +
        sibsp + parch + fare + embarked, titanic)
```

Note that we are not very much interested in the assessment of the model's predictive performance, but rather on understanding how the model yields its predictions. This is why we do not split the data into the training and testing subsets. Instead, the model is fitted to the entire dataset and will be examined on the same dataset.

4.2.2   Random forest model

As an alternative to the logistic regression model we consider a random forest model. Random forest modelling is known for good predictive performance, ability to grasp low-order variable interactions, and stability (Breiman, 2001a). To fit the model, we apply the `randomForest()` function, with default settings, from the package with the same name (Liaw and Wiener, 2002). In particular, we fit a model with the same set of explanatory variables as the logistic regression model (see Section 4.2.1). The results of the random forest model are stored in model-object `titanic_rf`.

```
library("randomForest")
set.seed(1313)
titanic_rf <- randomForest(survived ~ class + gender + age +
                sibsp + parch + fare + embarked, data = titanic)
```

4.2.3   Gradient boosting model

Additionally, we consider the gradient boosting model (Friedman, 2000). Tree-based boosting models are known for being able to accommodate higher-order interactions between variables. We use the same set of six explanatory variables as for the logistic regression model (see Section 4.2.1). To fit the gradient boosting model, we use function **gbm()** from the **gbm** package (Ridgeway, 2017). The results of the model are stored in model-object **titanic_gbm**.

```r
library("gbm")
set.seed(1313)
titanic_gbm <- gbm(survived == "yes" ~ class + gender + age +
                sibsp + parch + fare + embarked, data = titanic,
                n.trees = 15000, distribution = "bernoulli")
```

4.2.4 Support vector machine model

Finally, we also consider a support vector machine (SVM) model (Cortes and Vapnik, 1995). We use the C-classification mode. Again, we fit a model with the same set of explanatory variables as in the logistic regression model (see Section 4.2.1) To fit the model, we use function **svm()** from the **e1071** package (Meyer et al., 2019). The results of the model are stored in model-object **titanic_svm**.

```r
library("e1071")
titanic_svm <- svm(survived == "yes" ~ class + gender + age +
                sibsp + parch + fare + embarked, data = titanic,
                type = "C-classification", probability = TRUE)
```

4.2.5 Models' predictions

Let us now compare predictions that are obtained from the different models. In particular, we compute the predicted probability of survival for Johnny D, an 8-year-old boy who embarked in Southampton and travelled in the first class with no parents nor siblings, and with a ticket costing 72 pounds.

First, we create a data frame **johnny_d** that contains the data describing the passenger.

```r
johnny_d <- data.frame(
        class = factor("1st", levels = c("1st", "2nd", "3rd",
                    "deck crew", "engineering crew",
                    "restaurant staff", "victualling crew")),
        gender = factor("male", levels = c("female", "male")),
        age = 8, sibsp = 0, parch = 0, fare = 72,
        embarked = factor("Southampton", levels = c("Belfast",
                    "Cherbourg","Queenstown","Southampton")))
```

Subsequently, we use the generic function **predict()** to obtain the predicted probability of survival for the logistic regression model.

```r
(pred_lmr <- predict(titanic_lmr, johnny_d, type = "fitted"))
```

```
##        1
## 0.7677036
```

The predicted probability is equal to 0.77.

We do the same for the remaining three models.

```
(pred_rf <- predict(titanic_rf, johnny_d, type = "prob"))
```

```
##       no    yes
## 1 0.578 0.422
## attr(,"class")
## [1] "matrix" "array"  "votes"
```

```
(pred_gbm <- predict(titanic_gbm, johnny_d, type = "response",
                     n.trees = 15000))
```

```
## [1] 0.6632574
```

```
(pred_svm <- predict(titanic_svm, johnny_d, probability = TRUE))
```

```
##       1
## FALSE
## attr(,"probabilities")
##         FALSE      TRUE
## 1 0.7799685 0.2200315
## Levels: FALSE TRUE
```

As a result, we obtain the predicted probabilities of 0.42, 0.66, and 0.22 for the random forest, gradient boosting, and SVM models, respectively. The models lead to different probabilities. Thus, it might be of interest to understand the reason for the differences, as it could help us decide which of the predictions we might want to trust. We will investigate this issue in the subsequent chapters.

Note that, for some examples later in the book, we will use another observation (instance). We will call this passenger Henry.

```
henry <- data.frame(
    class = factor("1st", levels = c("1st", "2nd", "3rd",
                   "deck crew", "engineering crew",
                   "restaurant staff", "victualling crew")),
    gender = factor("male", levels = c("female", "male")),
    age = 47, sibsp = 0, parch = 0, fare = 25,
    embarked = factor("Cherbourg", levels = c("Belfast",
                      "Cherbourg","Queenstown","Southampton")))
```

For Henry, the predicted probability of survival is lower than for Johnny D.

```
predict(titanic_lmr, henry, type = "fitted")
```

```
##         1
## 0.4318245
```

```
predict(titanic_rf, henry, type = "prob")[,2]
```

```
## [1] 0.246
```

```
predict(titanic_gbm, henry, type = "response", n.trees = 15000)
```

```
## [1] 0.3073358
```

```
attr(predict(titanic_svm, henry, probability = TRUE),"probabilities")[,2]
```

```
## [1] 0.1767995
```

4.2.6 Models' explainers

Model-objects created with different libraries may have different internal structures. Thus, first, we have got to create an "explainer", i.e., an object that provides an uniform interface for different models. Toward this end, we use the `explain()` function from the DALEX package (Biecek, 2018). As it was mentioned in Section 3.1.2, there is only one argument that is required by the function, i.e., `model`. The argument is used to specify the model-object with the fitted form of the model. However, the function allows additional arguments that extend its functionalities. In particular, the list of arguments includes the following:

- `data`, a data frame or matrix providing data to which the model is to be applied; if not provided (`data = NULL` by default), the data are extracted from the model-object. Note that the data object should not, in principle, contain the dependent variable.
- `y`, observed values of the dependent variable corresponding to the data given in the `data` object; if not provided (`y = NULL` by default), the values are extracted from the model-object;
- `predict_function`, a function that returns prediction scores; if not specified (`predict_function = NULL` by default), then a default `predict()` function is used (note that this may lead to errors);
- `residual_function`, a function that returns model residuals; if not specified (`residual_function = NULL` by default), then model residuals defined in equation (2.1) are calculated;
- `verbose`, a logical argument (`verbose = TRUE` by default) indicating whether diagnostic messages are to be printed;
- `precalculate`, a logical argument (`precalculate = TRUE` by default) indicating whether predicted values and residuals are to be calculated when the explainer is created. Note that this will also happen if `verbose = TRUE`. To skip the calculations, both `verbose` and `precalculate` should be set to FALSE .
- `model_info`, a named list (with components `package`, `version`, and `type`) providing information about the model; if not specified (`model_info = NULL` by default), DALEX seeks for information on its own;
- `type`, information about the type of the model, either `"classification"` (for a binary dependent variable) or `"regression"` (for a continuous dependent

variable); if not specified (`type = NULL` by default), then the value of the argument is extracted from `model_info`;

- `label`, a unique name of the model; if not specified (`label = NULL` by default), then it is extracted from `class(model)`.

Application of function `explain()` provides an object of class `explainer`. It is a list of many components that include:

- `model`, the explained model;
- `data`, the data to which the model is applied;
- `y`, observed values of the dependent variable corresponding to `data`;
- `y_hat`, predictions obtained by applying `model` to `data`;
- `residuals`, residuals computed based on `y` and `y_hat`;
- `predict_function`, the function used to obtain the model's predictions;
- `residual_function`, the function used to obtain residuals;
- `class`, class/classes of the model;
- `label`, label of the model/explainer;
- `model_info`, a named list (with components `package`, `version`, and `type`) providing information about the model.

Thus, each explainer-object contains all elements needed to create a model explanation. The code below creates explainers for the models (see Sections 4.2.1–4.2.4) fitted to the Titanic data. Note that, in the `data` argument, we indicate the `titanic` data frame without the ninth column, i.e., without the *survived* variable. The variable is used in the `y` argument to explicitly define the binary dependent variable equal to 1 for survivors and 0 for passengers who did not survive.

```
titanic_lmr_exp <- explain(model = titanic_lmr,
                           data = titanic[, -9],
                           y = titanic$survived == "yes",
                           label = "Logistic Regression",
                           type = "classification")
titanic_rf_exp <- explain(model = titanic_rf,
                          data = titanic[, -9],
                          y = titanic$survived == "yes",
                          label = "Random Forest")
titanic_gbm_exp <- explain(model = titanic_gbm,
                           data = titanic[, -9],
                           y = titanic$survived == "yes",
                           label = "Generalized Boosted Regression")
titanic_svm_exp <- explain(model = titanic_svm,
                           data = titanic[, -9],
                           y = titanic$survived == "yes",
                           label = "Support Vector Machine")
```

4.2.7 List of model-objects

In the previous sections, we have built four predictive models for the Titanic dataset. The models will be used in the rest of the book to illustrate model-explanation methods and tools.

For the ease of reference, we summarize the models in Table 4.1. The binary model-objects can be downloaded by using the indicated `archivist` hooks (Biecek and Kosinski, 2017). By calling a function specified in the last column of the table, one can restore a selected model in its local R environment.

TABLE 4.1: Predictive models created for the `titanic` dataset. All models are fitted with following variables: *gender, age, class, sibsp, parch, fare, embarked*.

Model name / library	Link to this object
`titanic_lmr` `rms:: lmr` v.5.1.3	Get the model: `archivist:: aread("pbiecek/models/58b24")`.
`titanic_rf` `randomForest:: randomForest` v.4.6.14	Get the model: `archivist:: aread("pbiecek/models/4e0fc")`.
`titanic_gbm` `gbm:: gbm` v.2.1.5	Get the model: `archivist:: aread("pbiecek/models/b7078")`.
`titanic_svm` `e1071:: svm` v.1.7.3	Get the model: `archivist:: aread("pbiecek/models/9c27f")`.

Table 4.2 summarizes the data frames that will be used in examples in the subsequent chapters.

TABLE 4.2: Data frames created for the Titanic use-case. All frames include the following variables: *gender, age, class, embarked, country, fare, sibsp, parch*. The `titanic` data frame includes also the *survived* variable.

Description	Link to this object
`titanic` dataset with 2207 observations with imputed missing values	`archivist:: aread("pbiecek/models/27e5c")`
`johnny_d` 8-year-old boy from the 1st class without parents, paid 72 pounds, embarked in Southampton	`archivist:: aread("pbiecek/models/e3596")`

Description	Link to this object
henry 47-year-old male from the 1st class, travelled alone, paid 25 pounds, embarked in Cherbourg	`archivist::` `aread("pbiecek/models/a6538")`

4.3   Models for RMS Titanic, snippets for Python

Titanic data are provided in the `titanic` dataset, which is available in the `dalex` library. The values of the dependent binary variable are given in the `survived` column; the remaining columns give the values of the explanatory variables that are used to construct the classifiers.

The following instructions load the `titanic` dataset and split it into the dependent variable `y` and the explanatory variables `X`. Note that, for the purpose of this example, we do not divide the data into the training and testing sets. Instructions on how to deal with the situation when you want to analyze the model on data other than the training set will be presented in the subsequent chapters.

```
import dalex as dx
titanic = dx.datasets.load_titanic()
X = titanic.drop(columns='survived')
y = titanic.survived
```

Dataset `X` contains numeric variables with different ranges (for instance, *age* and *fare*) and categorical variables. Machine-learning algorithms in the `sklearn` library require data in a numeric form. Therefore, before modelling, we use a pipeline that performs data pre-processing. In particular, we scale the continuous variables (*age*, *fare*, *parch*, and *sibsp*) and one-hot-encode the categorical variables (*gender*, *class*, *embarked*).

```
from sklearn.preprocessing import StandardScaler, OneHotEncoder
from sklearn.compose import make_column_transformer
from sklearn.pipeline import make_pipeline

preprocess = make_column_transformer(
    (StandardScaler(), ['age', 'fare', 'parch', 'sibsp']),
    (OneHotEncoder(), ['gender', 'class', 'embarked']))
```

4.3.1   Logistic regression model

To fit the logistic regression model (see Section 4.2.1), we use the `LogisticRegression` algorithm from the `sklearn` library. By default, the implementation uses the ridge penalty, defined in (2.6). For this reason it is important to scale continuous variables like `age` and `fare`.

The fitted model is stored in object `titanic_lr`, which will be used in subsequent chapters.

```
from sklearn.linear_model import LogisticRegression

titanic_lr = make_pipeline(
    preprocess,
    LogisticRegression(penalty = '12'))
titanic_lr.fit(X, y)
```

4.3.2   Random forest model

To fit the random forest model (see Section 4.2.2), we use the `RandomForestClassifier` algorithm from the `sklearn` library. We use the default settings with trees not deeper than three levels, and the number of trees set to 500. The fitted model is stored in object `titanic_rf`.

```
from sklearn.ensemble import RandomForestClassifier

titanic_rf = make_pipeline(
    preprocess,
    RandomForestClassifier(max_depth = 3, n_estimators = 500))
titanic_rf.fit(X, y)
```

4.3.3   Gradient boosting model

To fit the gradient boosting model (see Section 4.2.3), we use the `GradientBoostingClassifier` algorithm from the `sklearn` library. We use the default settings, with the number of trees in the ensemble set to 100. The fitted model is stored in object `titanic_gbc`.

```
from sklearn.ensemble import GradientBoostingClassifier

titanic_gbc = make_pipeline(
    preprocess,
    GradientBoostingClassifier(n_estimators = 100))
titanic_gbc.fit(X, y)
```

4.3.4   Support vector machine model

Finally, to fit the SVM model with C-Support Vector Classification mode (see Section 4.2.4), we use the `SVC` algorithm from the `sklearn` library based on `libsvm`. The fitted model is stored in object `titanic_svm`.

```python
from sklearn.svm import SVC

titanic_svm = make_pipeline(
    preprocess,
    SVC(probability = True))
titanic_svm.fit(X, y)
```

4.3.5   Models' predictions

Let us now compare predictions that are obtained from the different models. In particular, we compute the predicted probability of survival for Johnny D, an 8-year-old boy who embarked in Southampton and travelled in the first class with no parents nor siblings, and with a ticket costing 72 pounds (see Section 4.2.5).

First, we create a data frame `johnny_d` that contains the data describing the passenger.

```python
import pandas as pd

johnny_d = pd.DataFrame({'gender': ['male'],
                         'age'     : [8],
                         'class'   : ['1st'],
                         'embarked': ['Southampton'],
                         'fare'    : [72],
                         'sibsp'   : [0],
                         'parch'   : [0]},
                        index = ['JohnnyD'])
```

Subsequently, we use the method `predict_proba()` to obtain the predicted probability of survival for the logistic regression model.

```python
titanic_lr.predict_proba(johnny_d)
# array([[0.35884528, 0.64115472]])
```

We do the same for the three remaining models.

```python
titanic_rf.predict_proba(johnny_d)
# array([[0.63028556, 0.36971444]])
titanic_gbc.predict_proba(johnny_d)
# array([[0.1567194, 0.8432806]])
titanic_svm.predict_proba(johnny_d)
# array([[0.78308146, 0.21691854]])
```

We also create data frame for passenger Henry (see Section 4.2.5) and compute his predicted probability of survival.

```python
henry = pd.DataFrame({'gender'   : ['male'],
                      'age'       : [47],
                      'class'     : ['1st'],
                      'embarked'  : ['Cherbourg'],
                      'fare'      : [25],
                      'sibsp'     : [0],
                      'parch'     : [0]},
                      index = ['Henry'])
titanic_lr.predict_proba(henry)
# array([[0.56798421 0.43201579]])
titanic_rf.predict_proba(henry)
# array([[0.69917845 0.30082155]])
titanic_gbc.predict_proba(henry)
# array([[0.78542886 0.21457114]])
titanic_svm.predict(henry)
# array([[0.81725832 0.18274168]])
```

4.3.6 Models' explainers

The Python-code examples shown above use functions from the **sklearn** library, which facilitates uniform working with models. However, we may want to, or have to, work with models built by using other libraries. To simplify the task, the **dalex** library wraps models in objects of class **Explainer** that contain, in a uniform way, all the functions necessary for working with models.

There is only one argument that is required by the **Explainer()** constructor, i.e., **model**. However, the constructor allows additional arguments that extend its functionalities. In particular, the list of arguments includes the following:

- **data**, a data frame or **numpy.ndarray** providing data to which the model is to be applied. It should be an object of the **pandas.DataFrame** class, otherwise it will be converted to **pandas.DataFrame**.
- **y**, values of the dependent variable/target variable corresponding to the data given in the **data** object;
- **predict_function**, a function that returns prediction scores; if not specified, then **dalex** will make a guess which function should be used (**predict()**, **predict_proba()**, or something else). Note that this function should work on **pandas.DataFrame** objects; if it works only on **numpy.ndarray** then an appropriate conversion should also be included in **predict_function**.
- **residual_function**, a function that returns model residuals;
- **label**, a unique name of the model;
- **model_class**, the class of actual model;
- **verbose**, a logical argument (**verbose = TRUE** by default) indicating whether diagnostic messages are to be printed;

- `model_type`, information about the type of the model, either `"classification"` (for a binary dependent variable) or `"regression"` (for a continuous dependent variable);
- `model_info`, a dictionary with additional information about the model.

Application of constructor `Explainer()` provides an object of class `Explainer`. It is an object with many components that include:

- `model`, the explained model;
- `data`, the data to which the model is applied;
- `y`, observed values of the dependent variable corresponding to `data`;
- `y_hat`, predictions obtained by applying `model` to `data`;
- `residuals`, residuals computed based on `y` and `y_hat`;
- `predict_function`, the function used to obtain the model's predictions;
- `residual_function`, the function used to obtain residuals;
- `class`, class/classes of the model;
- `label`, label of the model/explainer;
- `model_info`, a dictionary (with components `package`, `version`, and `type`) providing information about the model.

Thus, each explainer-object contains all elements needed to create a model explanation. The code below creates explainers for the models (see Sections 4.3.1–4.3.4) fitted to the Titanic data.

```
titanic_rf_exp = dx.Explainer(titanic_rf,
                 X, y, label = "Titanic RF Pipeline")
titanic_lr_exp = dx.Explainer(titanic_lr,
                 X, y, label = "Titanic LR Pipeline")
titanic_gbc_exp = dx.Explainer(titanic_gbc,
                 X, y, label = "Titanic GBC Pipeline")
titanic_svm_exp = dx.Explainer(titanic_svm,
                 X, y, label = "Titanic SVM Pipeline")
```

When an explainer is created, the specified model and data are tested for consistency. Diagnostic information is printed on the screen. The following output shows diagnostic information for the `titanic_rf` model.

```
Preparation of a~new explainer is initiated

  -> data               : 2207 rows 7 cols
  -> target variable    : Argument 'y' was converted to a~numpy.ndarray.
  -> target variable    : 2207 values
  -> model_class        : sklearn.pipeline.Pipeline (default)
  -> label              : Titanic RF Pipeline
  -> predict function   : <yhat_proba> will be used (default)
  -> predicted values   : min = 0.171, mean = 0.322, max = 0.893
  -> residual function  : difference between y and yhat (default)
  -> residuals          : min = -0.826, mean = 4.89e-05, max = 0.826
  -> model_info         : package sklearn
A new explainer has been created!
```

4.4 Apartment prices

Predicting house prices is a common exercise used in machine-learning courses. Various datasets for house prices are available at websites like Kaggle[1] or UCI Machine Learning Repository[2].

In this book, we will work with an interesting variant of this problem. The `apartments` dataset contains simulated data that match key characteristics of real apartments in Warsaw, the capital of Poland. However, the dataset is created in a way that two very different models, namely linear regression and random forest, offer almost exactly the same overall accuracy of predictions. The natural question is then: which model should we choose? We will show that model-explanation tools provide important insight into the key model characteristics and are helpful in model selection.

The dataset is available in the `DALEX` package in R and the `dalex` library in Python. It contains 1000 observations (apartments) and six variables:

- *m2.price*, apartment's price per square meter (in EUR), a numerical variable in the range of 1607–6595;
- *construction.year*, the year of construction of the block of flats in which the apartment is located, a numerical variable in the range of 1920–2010;
- *surface*, apartment's total surface in square meters, a numerical variable in the range of 20–150;
- *floor*, the floor at which the apartment is located (ground floor taken to be the first floor), a numerical integer variable with values ranging from 1 to 10;
- *no.rooms*, the total number of rooms, a numerical integer variable with values ranging from 1 to 6;
- *district*, a factor with 10 levels indicating the district of Warsaw where the apartment is located.

The first six rows of this dataset are presented in the table below.

m2.price	construction.year	surface	floor	no.rooms	district
5897	1953	25	3	1	Srodmiescie
1818	1992	143	9	5	Bielany
3643	1937	56	1	2	Praga
3517	1995	93	7	3	Ochota
3013	1992	144	6	5	Mokotow
5795	1926	61	6	2	Srodmiescie

---

[1]https://www.kaggle.com
[2]https://archive.ics.uci.edu

Models considered for this dataset will use *m2.price* as the (continuous) dependent variable. Models' predictions will be validated on a set of 9000 apartments included in data frame `apartments_test`.

Note that, usually, the training dataset is larger than the testing one. In this example, we deliberately use a small training set, so that model selection may be more challenging.

4.4.1   Data exploration

Note that `apartments` is an artificial dataset created to illustrate and explain differences between random forest and linear regression. Hence, the structure of the data, the form and strength of association between variables, plausibility of distributional assumptions, etc., is less problematic than in a real-life dataset. In fact, all these characteristics of the data are known. Nevertheless, we present some data exploration below to illustrate the important aspects of the data.

The variable of interest is *m2.price*, the price per square meter. The histogram presented in Figure 4.6 indicates that the distribution of the variable is slightly skewed to the right.

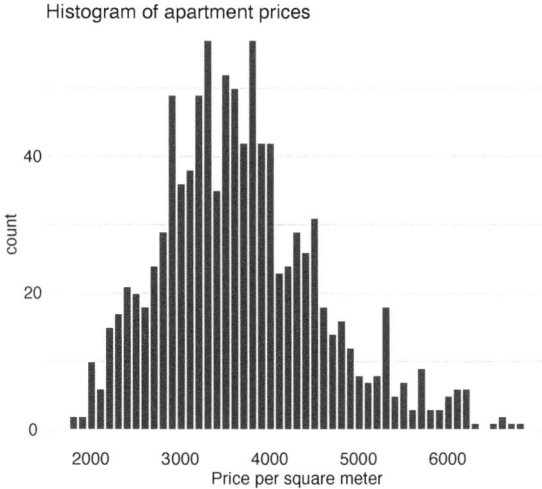

FIGURE 4.6 Distribution of the price per square meter in the apartment-prices data.

Figure 4.7 suggests (possibly) a non-linear relationship between *construction.year* and *m2.price* and a linear relation between *surface* and *m2.price*.

Figure 4.8 indicates that the relationship between *floor* and *m2.price* is also close to linear, as well as is the association between *no.rooms* and *m2.price* .

Figure 4.9 shows that *surface* and *number of rooms* are positively associated and that prices depend on the district. In particular, box plots in Figure 4.9

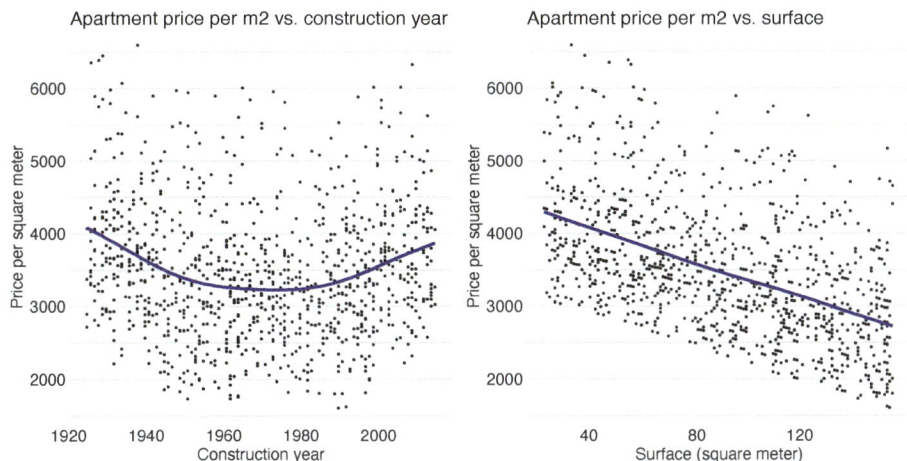

FIGURE 4.7 Apartment-prices data. Price per square meter vs. year of construction (left-hand-side panel) and vs. surface (right-hand-side panel).

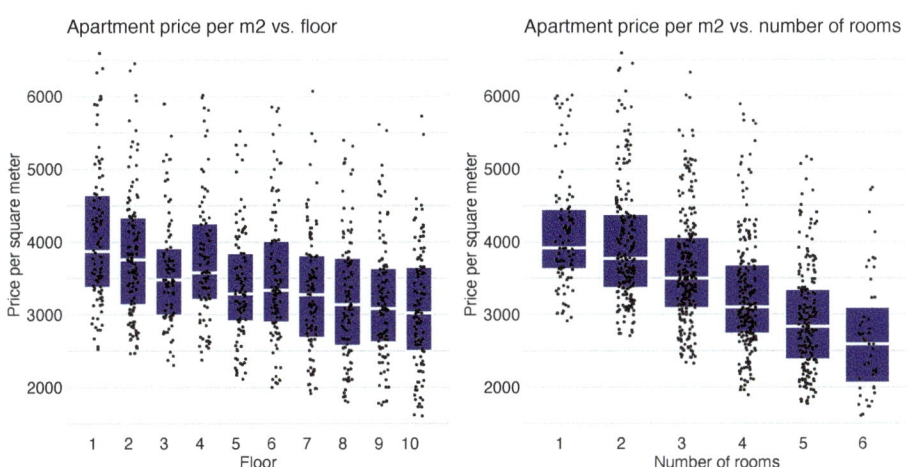

FIGURE 4.8 Apartment-prices data. Price per square meter vs. floor (left-hand-side panel) and vs. number of rooms (right-hand-side panel).

indicate that the highest prices per square meter are observed in Srodmiescie (Downtown).

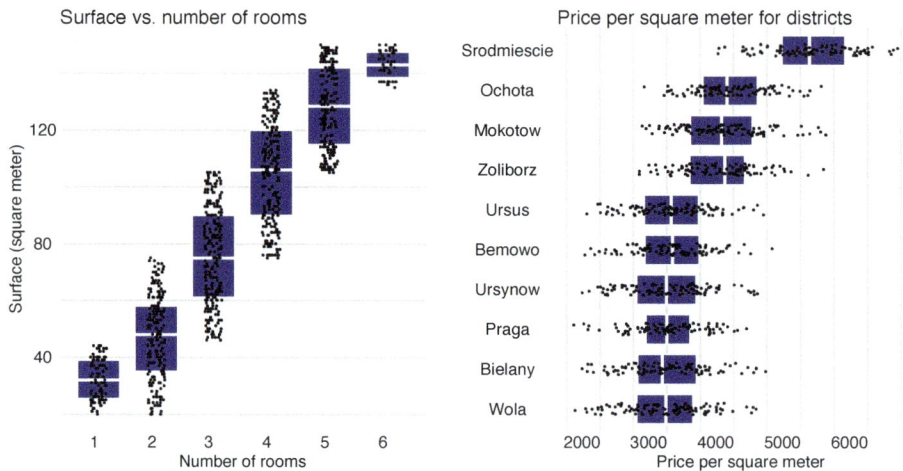

FIGURE 4.9 Apartment-prices data. Surface vs. number of rooms (left-hand-side panel) and price per square meter for different districts (right-hand-side panel).

4.5 Models for apartment prices, snippets for R

4.5.1 Linear regression model

The dependent variable of interest, *m2.price*, is continuous. Thus, a natural choice to build a predictive model is linear regression. We treat all the other variables in the `apartments` data frame as explanatory and include them in the model. To fit the model, we apply the `lm()` function. The results of the model are stored in model-object `apartments_lm`.

```
library("DALEX")
apartments_lm <- lm(m2.price ~ ., data = apartments)
anova(apartments_lm)
```

```
## Analysis of Variance Table
##
## Response: m2.price
##                     Df     Sum Sq    Mean Sq F value    Pr(>F)
## construction.year    1    2629802    2629802   33.233 1.093e-08 ***
## surface              1  207840733  207840733 2626.541 < 2.2e-16 ***
## floor                1   79823027   79823027 1008.746 < 2.2e-16 ***
```

```
## no.rooms              1     956996     956996   12.094  0.000528 ***
## district             9 451993980   50221553  634.664 < 2.2e-16 ***
## Residuals          986  78023123      79131
## ---
## Signif. codes:  0 '***' 0.001 '**' 0.01 '*' 0.05 '.' 0.1 ' ' 1
```

4.5.2   Random forest model

As an alternative to linear regression, we consider a random forest model.
Again, we treat all the variables in the `apartments` data frame other than
m2.price as explanatory and include them in the model. To fit the model, we
apply the `randomForest()` function, with default settings, from the package
with the same name (Liaw and Wiener, 2002). The results of the model are
stored in model-object `apartments_rf`.

```
library("randomForest")
set.seed(72)
apartments_rf <- randomForest(m2.price ~ ., data = apartments)
```

4.5.3   Support vector machine model

Finally, we consider an SVM model, with all the variables in the `apartments`
data frame other than *m2.price* treated as explanatory. To fit the model, we use
the `svm()` function, with default settings, from package e1071 (Meyer et al.,
2019). The results of the model are stored in model-object `apartments_svm`.

```
library("e1071")
apartments_svm <- svm(m2.price ~ construction.year + surface + floor +
        no.rooms + district, data = apartments)
```

4.5.4   Models' predictions

The `predict()` function calculates predictions for a specific model. In the
example below, we use model-objects `apartments_lm`, `apartments_rf`, and
`apartments_svm`, to calculate predictions for prices of the apartments from
the `apartments_test` data frame. Note that, for brevity's sake, we compute
the predictions only for the first six observations from the data frame.

The actual prices for the first six observations from `apartments_test` are
provided below.

```
apartments_test$m2.price[1:6]
```

```
## [1] 4644 3082 2498 2735 2781 2936
```

Predicted prices for the linear regression model are as follows:

```
predict(apartments_lm, apartments_test[1:6,])
```

```
##      1001      1002      1003      1004      1005      1006
## 4820.009 3292.678 2717.910 2922.751 2974.086 2527.043
```

Predicted prices for the random forest model take the following values:

```
predict(apartments_rf, apartments_test[1:6,])
```

```
##      1001      1002      1003      1004      1005      1006
## 4214.084 3178.061 2695.787 2744.775 2951.069 2999.450
```

Predicted prices for the SVM model are as follows:

```
predict(apartments_svm, apartments_test[1:6,])
```

```
##      1001      1002      1003      1004      1005      1006
## 4590.076 3012.044 2369.748 2712.456 2681.777 2750.904
```

By using the code presented below, we summarize the predictive performance of the linear regression and random forest models by computing the square root of the mean-squared-error (RMSE). For a "perfect" predictive model, which would predict all observations exactly, RMSE should be equal to 0. More information about RMSE can be found in Section 15.3.1.

```
predicted_apartments_lm <- predict(apartments_lm, apartments_test)
sqrt(mean((predicted_apartments_lm - apartments_test$m2.price)^2))
```

```
## [1] 283.0865
```

```
predicted_apartments_rf <- predict(apartments_rf, apartments_test)
sqrt(mean((predicted_apartments_rf - apartments_test$m2.price)^2))
```

```
## [1] 282.9519
```

For the random forest model, RMSE is equal to 283. It is almost identical to the RMSE for the linear regression model, which is equal to 283.1. Thus, the question we may face is: should we choose the more complex but flexible random forest model, or the simpler and easier to interpret linear regression model? In the subsequent chapters, we will try to provide an answer to this question. In particular, we will show that a proper model exploration may help to discover weak and strong sides of any of the models and, in consequence, allow the creation of a new model, with better performance than either of the two.

4.5.5 Models' explainers

The code presented below creates explainers for the models (see Sections 4.5.1–4.5.3) fitted to the apartment-prices data. Note that we use the `apartments_test` data frame without the first column, i.e., the *m2.price* variable, in the `data` argument. This will be the dataset to which the model will be applied (see Section 4.2.6). The *m2.price* variable is explicitly specified as the dependent variable in the y argument (see Section 4.2.6).

```
apartments_lm_exp <- explain(model = apartments_lm,
                             data = apartments_test[,-1],
                             y = apartments_test$m2.price,
                             label = "Linear Regression")
apartments_rf_exp <- explain(model = apartments_rf,
                             data = apartments_test[,-1],
                             y = apartments_test$m2.price,
                             label = "Random Forest")
apartments_svm_exp <- explain(model = apartments_svm,
                             data = apartments_test[,-1],
                             y = apartments_test$m2.price,
                             label = "Support Vector Machine")
```

4.5.6   List of model-objects

In Sections 4.5.1–4.5.3, we have built three predictive models for the `apartments` dataset. The models will be used in the rest of the book to illustrate the model-explanation methods and tools.

For the ease of reference, we summarize the models in Table 4.3. The binary model-objects can be downloaded by using the indicated `archivist` hooks (Biecek and Kosinski, 2017). By calling a function specified in the last column of the table, one can restore a selected model in a local R environment.

TABLE 4.3: Predictive models created for the dataset Apartment prices. All models are fitted by using *construction.year, surface, floor, no.rooms*, and *district* as explanatory variables.

Model name / library	Link to this object
`apartments_lm` `stats:: lm v.3.5.3`	Get the model: `archivist::` `aread("pbiecek/models/55f19")`.
`apartments_rf` `randomForest:: randomForest v.4.6.14`	Get the model: `archivist::` `aread("pbiecek/models/fe7a5")`.
`apartments_svm` `e1071:: svm v.1.7.3`	Get the model: `archivist::` `aread("pbiecek/models/d2ca0")`.

4.6   Models for apartment prices, snippets for Python

Apartment-prices data are provided in the `apartments` dataset, which is available in the `dalex` library. The values of the continuous dependent variable are given in the `m2_price` column; the remaining columns give the values of the explanatory variables that are used to construct the predictive models.

The following instructions load the **apartments** dataset and split it into the dependent variable y and the explanatory variables X.

```
import dalex as dx
apartments = dx.datasets.load_apartments()
X = apartments.drop(columns='m2_price')
y = apartments['m2_price']
```

Dataset X contains numeric variables with different ranges (for instance, *surface* and *no.rooms*) and categorical variables (*district*). Machine-learning algorithms in the **sklearn** library require data in a numeric form. Therefore, before modelling, we use a pipeline that performs data pre-processing. In particular, we scale the continuous variables (*construction.year*, *surface*, *floor*, and *no.rooms*) and one-hot-encode the categorical variables (*district*).

```
from sklearn.preprocessing import StandardScaler, OneHotEncoder
from sklearn.compose import make_column_transformer
from sklearn.pipeline import make_pipeline

preprocess = make_column_transformer(
    (StandardScaler(), ['construction_year', 'surface', 'floor',
        'no_rooms']),
    (OneHotEncoder(), ['district']))
```

4.6.1   Linear regression model

To fit the linear regression model (see Section 4.5.1), we use the `LinearRegression` algorithm from the **sklearn** library. The fitted model is stored in object **apartments_lm**, which will be used in subsequent chapters.

```
from sklearn.linear_model import LinearRegression

apartments_lm = make_pipeline(
    preprocess,
    LinearRegression())
apartments_lm.fit(X, y)
```

4.6.2   Random forest model

To fit the random forest model (see Section 4.5.2), we use the `RandomForestRegressor` algorithm from the **sklearn** library. We apply the default settings with trees not deeper than three levels and the number of trees in the random forest set to 500. The fitted model is stored in object **apartments_rf** for purpose of illustrations in subsequent chapters.

```
from sklearn.ensemble import RandomForestRegressor

apartments_rf = make_pipeline(
    preprocess,
    RandomForestRegressor(max_depth = 3, n_estimators = 500))
apartments_rf.fit(X, y)
```

4.6.3   Support vector machine model

Finally, to fit the SVM model (see Section 4.5.3), we use the `SVR` algorithm from the `sklearn` library. The fitted model is stored in object `apartments_svm`, which will be used in subsequent chapters.

```
from sklearn.svm import SVR

apartments_svm = make_pipeline(
    preprocess,
    SVR())
apartments_svm.fit(X, y)
```

4.6.4   Models' predictions

Let us now compare predictions that are obtained from the different models for the `apartments_test` data. In the code below, we use the `predict()` method to obtain the predicted price per square meter for the linear regression model.

```
apartments_test = dx.datasets.load_apartments_test()
apartments_test = apartments_test.drop(columns='m2_price')

apartments_lm.predict(apartments_test)
# array([4820.00943156, 3292.67756996, 2717.90972101,..., 4836.44370353,
#        3191.69063189, 5157.93680175])
```

In a similar way, we obtain the predictions for the two remaining models.

```
apartments_rf.predict(apartments_test)
# array([4708, 3819, 2273,..., 4708, 4336, 4916])

apartments_svm.predict(apartments_test)
# array([3344.48570564, 3323.01215313, 3321.97053977,..., 3353.19750146,
#        3383.51743883, 3376.31070911])
```

4.6.5   Models' explainers

The Python-code examples presented for the models for the apartment-prices dataset use functions from the `sklearn` library, which facilitates uniform working with models. However, we may want to, or have to, work with models built by using other libraries. To simplify the task, the **dalex** library wraps models in objects of class **Explainer** that contain, in a uniform way, all the functions necessary for working with models (see Section 4.3.6). The code below creates explainer-objects for the models (see Sections 4.6.1–4.6.3) fitted to the apartment-prices data.

```
apartments_lm_exp = dx.Explainer(apartments_lm, X, y,
                      label = "Apartments LM Pipeline")
apartments_rf_exp = dx.Explainer(apartments_rf, X, y,
                      label = "Apartments RF Pipeline")
apartments_svm_exp = dx.Explainer(apartments_svm, X, y,
                      label = "Apartments SVM Pipeline")
```

When an explainer is created, the specified model and data are tested for consistency. Diagnostic information is printed on the screen. The following output shows diagnostic information for the **apartments_lm** model.

```
Preparation of a~new explainer is initiated

  -> data               : 1000 rows 5 cols
  -> target variable    : Argument 'y' converted to a~numpy.ndarray.
  -> target variable    : 1000 values
  -> model_class        : sklearn.pipeline.Pipeline (default)
  -> label              : Apartments LM Pipeline
  -> predict function   : <yhat at 0x117090840> will be used (default)
  -> predicted values   : min = 1.78e+03, mean = 3.49e+03, max = 6.18e+03
  -> residual function  : difference between y and yhat (default)
  -> residuals          : min = -2.47e+02, mean = 2.06e-13, max = 4.69e+02
  -> model_info         : package sklearn

A new explainer has been created!
```

Part II

Instance Level

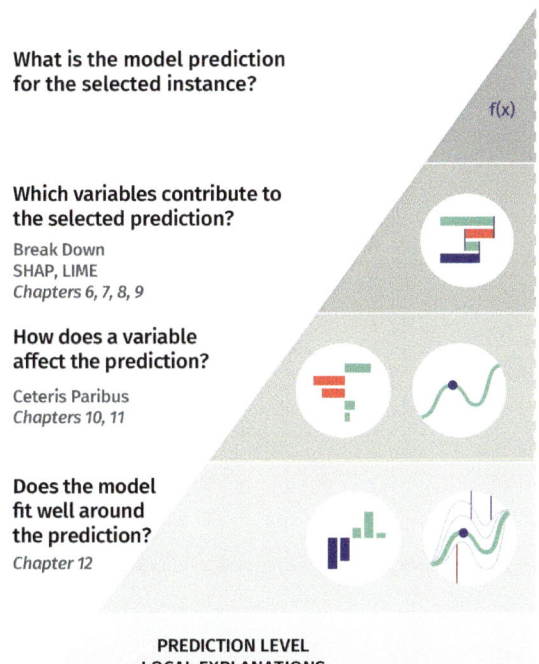

What is the model prediction
for the selected instance?

f(x)

Which variables contribute to
the selected prediction?

Break Down
SHAP, LIME
Chapters 6, 7, 8, 9

How does a variable
affect the prediction?

Ceteris Paribus
Chapters 10, 11

Does the model
fit well around
the prediction?
Chapter 12

PREDICTION LEVEL
LOCAL EXPLANATIONS

5

Introduction to Instance-level Exploration

Instance-level exploration methods help us understand how a model yields a prediction for a particular single observation. We may consider the following situations as examples:

- We may want to evaluate effects of explanatory variables on the model's predictions. For instance, we may be interested in predicting the risk of heart attack based on a person's age, sex, and smoking habits. A model may be used to construct a score (for instance, a linear combination of the explanatory variables representing age, sex, and smoking habits) that could be used for the purposes of prediction. For a particular patient, we may want to learn how much do the different variables contribute to the score?
- We may want to understand how would the model's predictions change if values of some of the explanatory variables changed? For instance, what would be the predicted risk of heart attack if the patient cut the number of cigarettes smoked per day by half?
- We may discover that the model is providing incorrect predictions, and we may want to find the reason. For instance, a patient with a very low risk-score experienced a heart attack. What has driven the wrong prediction?

In this part of the book, we describe the most popular approaches to instance-level exploration. They can be divided into three classes:

- One approach is to analyze how does the model's prediction for a particular instance differ from the average prediction and how can the difference be distributed among explanatory variables? This method is often called the "variable attributions" approach. An example is provided in panel A of Figure 5.1. Chapters 6-8 present various methods for implementing this approach.
- Another approach uses the interpretation of the model as a function and investigates the local behaviour of this function around the point (observation) of interest $\underline{x}_*$. In particular, we analyze the curvature of the model response (prediction) surface around $\underline{x}_*$. In case of a black-box model, we may approximate it with a simpler glass-box model around $\underline{x}_*$. An example is provided in panel B of Figure 5.1. Chapter 9 presents the Local Interpretable Model-agnostic Explanations (LIME) method that exploits the concept of a "local model".
- Yet another approach is to investigate how does the model's prediction change if the value of a single explanatory variable changes? The approach

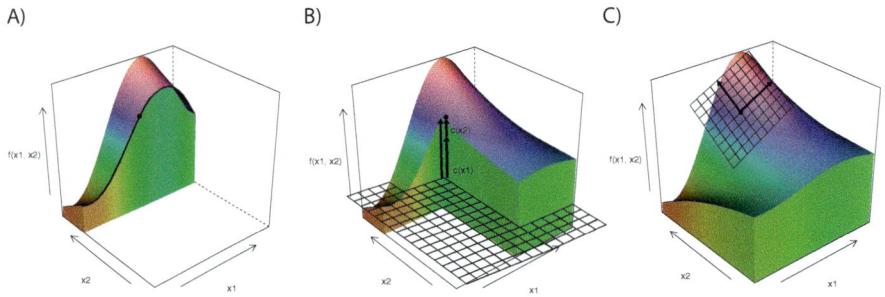

FIGURE 5.1 Illustration of different approaches to instance-level exploration. The plots present response (prediction) surface for a (black-box) model that is a function of two explanatory variables. We are interested in understanding the model response (prediction) at a single point (observation). Panel A illustrates the concept of variable attributions. The additive effect of each variable shows how does the prediction for the particular observation differ from the average. Panel B illustrates the concept of explanations through local models. A simpler glass-box model is used to approximate the black-box model around the point (observation) of interest. It describes the local behaviour of the model. Panel C presents a "What-if" analysis with a ceteris-paribus profile. The profile shows the model response (prediction) as a function of a single explanatory variable while keeping the values of all other explanatory variables fixed.

is useful in the so-called "What-if" analyses. In particular, we can construct plots presenting the change in model-based predictions induced by a change of a single explanatory variable. Such plots are usually called ceteris-paribus (CP) profiles. An example is provided in panel C in Figure 5.1. Chapters 10-12 introduce the CP profiles and methods based on them.

Each method has its own merits and limitations. They are briefly discussed in the corresponding chapters. Chapter 13 offers a comparison of the methods.

6

Break-down Plots for Additive Attributions

6.1 Introduction

Probably the most commonly asked question when trying to understand a model's prediction for a single observation is: *which variables contribute to this result the most?* There is no single best approach that can be used to answer this question. In this chapter, we introduce break-down (BD) plots, which offer a possible solution. The plots can be used to present "variable attributions", i.e., the decomposition of the model's prediction into contributions that can be attributed to different explanatory variables. Note that the method is similar to the `EXPLAIN` algorithm introduced by Robnik-Šikonja and Kononenko (2008) and implemented in the `ExplainPrediction` package (Robnik-Šikonja, 2018).

6.2 Intuition

As mentioned in Section 2.5, we assume that prediction $f(\underline{x})$ is an approximation of the expected value of the dependent variable Y given values of explanatory variables $\underline{x}$. The underlying idea of BD plots is to capture the contribution of an explanatory variable to the model's prediction by computing the shift in the expected value of Y, while fixing the values of other variables.

This idea is illustrated in Figure 6.1. Consider an example related to the prediction obtained for the random forest model `model_rf` for the Titanic data (see Section 4.2.2). We are interested in the probability of survival for Johnny D, an 8-year-old passenger travelling in the first class (see Section 4.2.5). Panel A of Figure 6.1 shows the distribution of the model's predictions for observations from the Titanic dataset. In particular, the violin plot in the row marked "all data" summarizes the distribution of predictions for all 2207 observations from the dataset. The red dot indicates the mean value that can be interpreted as an estimate of the expected value of the model's predictions over the distribution of all explanatory variables. In this example, the mean value is equal to 23.5%.

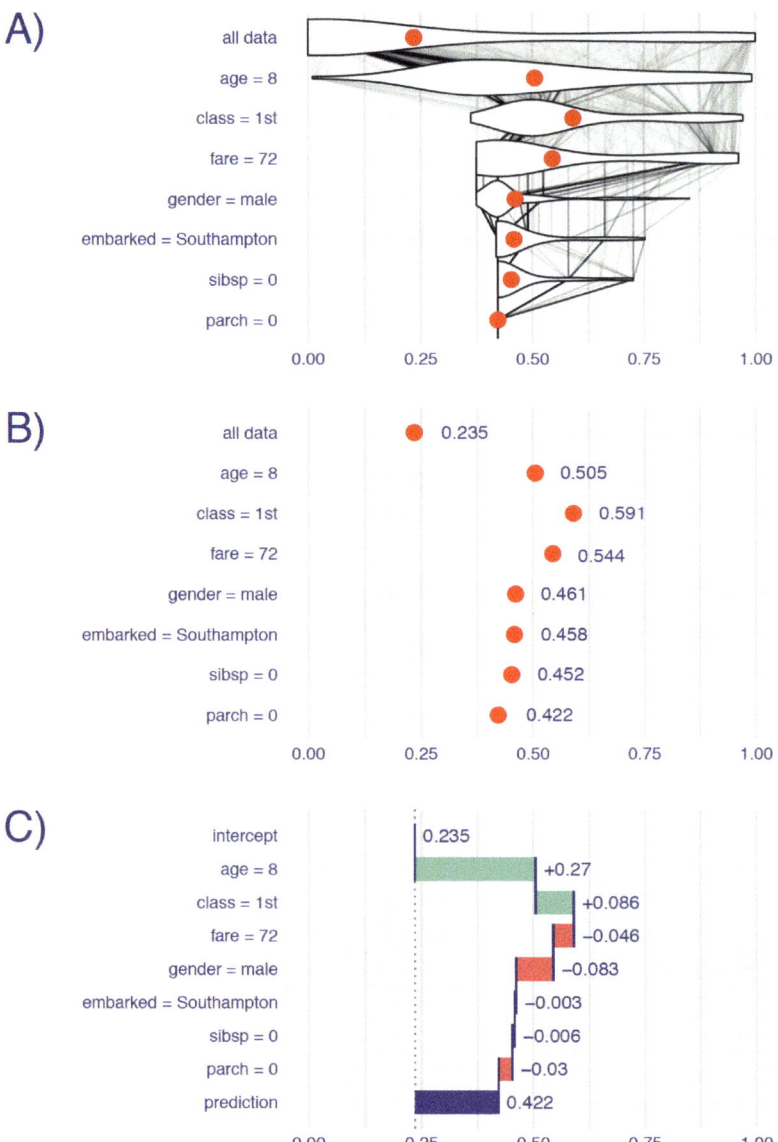

FIGURE 6.1 Break-down plots show how the contributions attributed to individual explanatory variables change the mean model's prediction to yield the actual prediction for a particular single instance (observation). Panel A) The first row shows the distribution and the mean value (red dot) of the model's predictions for all data. The next rows show the distribution and the mean value of the predictions when fixing values of subsequent explanatory variables. The last row shows the prediction for the particular instance of interest. B) Red dots indicate the mean predictions from panel A. C) The green and red bars indicate, respectively, positive and negative changes in the mean predictions (contributions attributed to explanatory variables).

To evaluate the contribution of individual explanatory variables to this particular single-instance prediction, we investigate the changes in the model's predictions when fixing the values of consecutive variables. For instance, the violin plot in the row marked "age=8" in panel A of Figure 6.1 summarizes the distribution of the predictions obtained when the *age* variable takes the value "8 years", as for Johnny D. Again, the red dot indicates the mean of the predictions, and it can be interpreted as an estimate of the expected value of the predictions over the distribution of all explanatory variables other than *age*. The violin plot in the "class=1st" row describes the distribution and the mean value of predictions with the values of variables *age* and *class* set to "8 years" and "1st class", respectively. Subsequent rows contain similar information for other explanatory variables included in the random forest model. In the last row, all explanatory variables are fixed at the values describing Johnny D. Hence, the last row contains only one point, the red dot, which corresponds to the model's prediction, i.e., survival probability, for Johnny D.

The thin grey lines in panel A of Figure 6.1 show the change of predictions for different individuals when the value of a particular explanatory variable is being replaced by the value indicated in the name of the row. For instance, the lines between the first and the second row indicate that fixing the value of the *age* variable to "8 years" has a different effect for different individuals. For some individuals (most likely, passengers that are 8 years old) the model's prediction does not change at all. For others, the predicted value increases (probably for the passengers older than 8 years) or decreases (most likely for the passengers younger than 8 years).

Eventually, however, we may be interested in the mean predictions, or even only in the changes of the means. Thus, simplified plots, similar to those shown in panels B and C of Figure 6.1, may be of interest. Note that, in panel C, the row marked "intercept" presents the overall mean value (0.235) of predictions for the entire dataset. Consecutive rows present changes in the mean prediction induced by fixing the value of a particular explanatory variable. Positive changes are indicated with green bars, while negative differences are indicated with red bars. The last row, marked "prediction", contains the sum of the overall mean value and the changes, i.e., the predicted value of survival probability for Johnny D, indicated by the blue bar.

What can be learned from BD plots as those presented in Figure 6.1? The plots offer a summary of the effects of particular explanatory variables on a model's predictions.

From Figure 6.1 we can conclude, for instance, that the mean prediction for the random forest model for the Titanic dataset is equal to 23.5%. This is the predicted probability of survival averaged over all people on Titanic. Note that it is not the percentage of individuals that survived, but the mean model-prediction. Thus, for a different model, we would most likely obtain a different mean value.

The model's prediction for Johnny D is equal to 42.2%. It is much higher than the mean prediction. The two explanatory variables that influence this prediction the most are *class* (with the value "1st") and *age* (with the value equal to 8). By fixing the values of these two variables, we add 35.6 percentage points to the mean prediction. All other explanatory variables have smaller effects, and they actually reduce the increase in the predicted value induced by *class* and *age*. For instance, *gender* (Johnny D was a boy) reduces the predicted survival probability by about 8.3 percentage points.

It is worth noting that the part of the prediction attributed to an explanatory variable depends not only on the variable but also on the considered value. For instance, in the example presented in Figure 6.1, the effect of the *embarked* harbour is very small. This may be due to the fact that the variable is not very important for prediction. However, it is also possible that the variable is important, but the effect of the value considered for the particular instance (Johnny D, who embarked Titanic in Southampton) may be close to the mean, as compared to all other possible values of the variable.

It is also worth mentioning that, for models that include interactions, the part of the prediction attributed to a variable depends on the order in which one sets the values of the explanatory variables. Note that the interactions do not have to be explicitly specified in the model structure as it is the case of, for instance, linear-regression models. They may also emerge as a result of fitting to the data a flexible model like, for instance, a regression tree.

To illustrate the point, Figure 6.2 presents an example of a random forest model with only three explanatory variables fitted to the Titanic data. Subsequently, we focus on the model's prediction for a 2-year old boy that travelled in the second class. The predicted probability of survival is equal to 0.964, more than a double of the mean model-prediction of 0.407. We would like to understand which explanatory variables drive this prediction. Two possible explanations are illustrated in Figure 6.2.

Explanation 1:

We first consider the explanatory variables *gender*, *class*, and *age*, in that order. Figure 6.2 indicates negative contributions for the first two variables and a positive contribution for the third one. Thus, the fact that the passenger was a boy decreases the chances of survival, as compared to the mean model-prediction. He travelled in the second class, which further lowers the probability of survival. However, as the boy was very young, this substantially increases the odds of surviving. This last conclusion is the result of the fact that most passengers in the second class were adults; therefore, a kid from the second class had higher chances of survival.

Explanation 2:

We now consider the following order of explanatory variables: *gender*, *age*, and *class*. Figure 6.2 indicates a positive contribution of *class*, unlike in the first explanation. Again, the fact that the passenger was a boy decreases the chances of survival, as compared to the mean model-prediction. However, he was very young, and this increases the probability of survival as compared to adult men. Finally, the fact that the boy travelled in the second class increases the chance even further. This last conclusion stems from the fact that most kids travelled in the third class; thus, being a child in the second class would increase chances of survival.

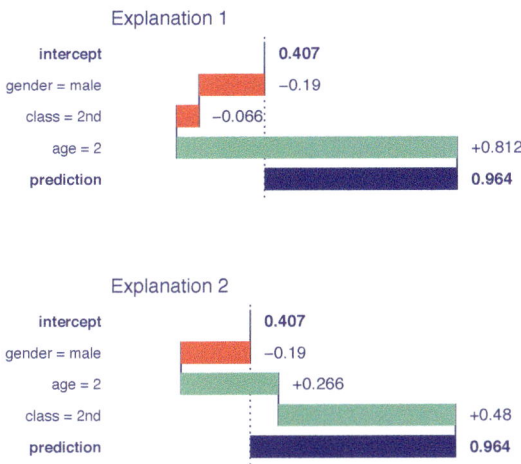

FIGURE 6.2 An illustration of the order-dependence of variable attributions. Two break-down plots for the same observation for a random forest model for the Titanic dataset. The contribution attributed to class is negative in the plot at the top and positive in the one at the bottom. The difference is due to the difference in the ordering of explanatory variables used to construct the plots (as seen in the labelling of the rows).

6.3   Method

In this section, we introduce more formally the method of variable attribution. We first focus on linear models, because their simple and additive structure allows building intuition. Then we consider a more general case.

6.3.1   Break-down for linear models

Assume the classical linear-regression model for dependent variable Y with p explanatory variables, the values of which are collected in vector $\underline{x}$, and vector β of p corresponding coefficients. Note that we separately consider β^0, which is the intercept. Prediction for Y is given by the expected value of Y conditional on $\underline{x}$. In particular, the expected value is given by the following linear combination:

$$E_Y(Y|\underline{x}) = f(\underline{x}) = \beta^0 + \underline{x}'\beta. \tag{6.1}$$

Assume that we select a vector of values of explanatory variables $\underline{x}_* \in \mathcal{R}^p$. We are interested in the contribution of the j-th explanatory variable to model's prediction $f(\underline{x}_*)$ for a single observation described by $\underline{x}_*$.

A possible approach to evaluate the contribution is to measure how much the expected value of Y changes after conditioning on x_*^j. Using the notation $x_*^{j|=X^j}$ (see Section 2.3) to indicate that we treat the value of the j-th coordinate as a random variable X^j, we can thus define

$$v(j, \underline{x}_*) = E_Y(Y|\underline{x}_*) - E_{X^j}\left[E_Y\left\{Y|\underline{x}_*^{j|=X^j}\right\}\right] = f(x_*) - E_{X^j}\left\{f\left(\underline{x}_*^{j|=X^j}\right)\right\}, \tag{6.2}$$

where $v(j, \underline{x}_*)$ is the *variable-importance measure* for the j-th explanatory variable evaluated at $\underline{x}_*$ and the last expected value on the right-hand side of (6.2) is taken over the distribution of the variable (treated as random). For the linear-regression model (6.1), and if the explanatory variables are independent, $v(j, \underline{x}_*)$ can be expressed as follows:

$$v(j, \underline{x}_*) = \beta^0 + \underline{x}_*'\underline{\beta} - E_{X^j}\left\{\beta^0 + \left(\underline{x}_*^{j|=X^j}\right)'\underline{\beta}\right\} = \beta^j\left\{x_*^j - E_{X^j}(X^j)\right\}. \tag{6.3}$$

Using (6.3), the linear-regression prediction (6.1) may be re-expressed in the following way:

$$f(\underline{x}_*) = \{\beta^0 + \beta^1 E_{X^1}(X^1) + \ldots + \beta^p E_{X^p}(X^p)\} +$$
$$[\{x_*^1 - E_{X^1}(X^1)\}\beta^1 + \ldots + \{x_*^p - E_{X^p}(X^p)\}\beta^p]$$
$$\equiv (mean\ prediction) + \sum_{j=1}^{p} v(j, \underline{x}_*). \tag{6.4}$$

Thus, the contributions of the explanatory variables $v(j, \underline{x}_*)$ sum up to the difference between the model's prediction for $\underline{x}_*$ and the mean prediction.

In practice, given a dataset, the expected value $E_{X^j}(X^j)$ can be estimated by the sample mean $\bar{x}^j$. This leads to

$$v(j, \underline{x}_*) = \beta^j (x_*^j - \bar{x}^j). \tag{6.5}$$

Obviously, the sample mean $\bar{x}^j$ is an estimator of the expected value $E_{X^j}(X^j)$, calculated using a dataset. For the sake of simplicity, we do not emphasize this difference in the notation. Also, we ignore the fact that, in practice, we never know the true model coefficients and use their estimates instead. We are also silent about the fact that, usually, explanatory variables are not independent. We needed this simplified example just to build our intuition.

6.3.2 Break-down for a general case

Again, let $v(j, \underline{x}_*)$ denote the variable-importance measure of the j-th variable and instance $\underline{x}_*$, i.e., the contribution of the j-th variable to the model's prediction at $\underline{x}_*$.

We would like the sum of the $v(j, \underline{x}_*)$ for all explanatory variables to be equal to the instance prediction. This property is called *local accuracy*. Thus, we want that

$$f(\underline{x}_*) = v_0 + \sum_{j=1}^{p} v(j, \underline{x}_*), \tag{6.6}$$

where v_0 denotes the mean model-prediction. Denote by $\underline{X}$ the vector of random values of explanatory variables. If we rewrite equation (6.6) as follows:

$$E_{\underline{X}}\{f(\underline{X})|X^1 = x_*^1, \ldots, X^p = x_*^p\} = E_{\underline{X}}\{f(\underline{X})\} + \sum_{j=1}^{p} v(j, \underline{x}_*), \tag{6.7}$$

then a natural proposal for $v(j, \underline{x}_*)$ is

$$\begin{aligned} v(j, \underline{x}_*) = &E_{\underline{X}}\{f(\underline{X})|X^1 = x_*^1, \ldots, X^j = x_*^j\} - \\ &E_{\underline{X}}\{f(\underline{X})|X^1 = x_*^1, \ldots, X^{j-1} = x_*^{j-1}\}. \end{aligned} \tag{6.8}$$

In other words, the contribution of the j-th variable is the difference between the expected value of the model's prediction conditional on setting the values of the first j variables equal to their values in $\underline{x}_*$ and the expected value conditional on setting the values of the first $j-1$ variables equal to their values in $\underline{x}_*$.

Note that the definition does imply the dependence of $v(j, \underline{x}_*)$ on the order of the explanatory variables that is reflected in their indices (superscripts).

To consider more general cases, let J denote a subset of K indices $(K \leq p)$ from $\{1, 2, \ldots, p\}$, i.e., $J = \{j_1, j_2, \ldots, j_K\}$, where each $j_k \in \{1, 2, \ldots, p\}$. Furthermore, let L denote another subset of M indices $(M \leq p - K)$ from $\{1, 2, \ldots, p\}$, distinct from J. That is, $L = \{l_1, l_2, \ldots, l_M\}$, where each $l_m \in \{1, 2, \ldots, p\}$ and $J \cap L = \emptyset$. Let us define now

$$\Delta^{L|J}(\underline{x}_*) \equiv E_{\underline{X}}\{f(\underline{X})|X^{l_1} = x_*^{l_1}, \ldots, X^{l_M} = x_*^{l_M}, X^{j_1} = x_*^{j_1}, \ldots, X^{j_K} = x_*^{j_K}\} -$$
$$E_{\underline{X}}\{f(\underline{X})|X^{j_1} = x_*^{j_1}, \ldots, X^{j_K} = x_*^{j_K}\}. \tag{6.9}$$

In other words, $\Delta^{L|J}(\underline{x}_*)$ is the change between the expected model-prediction, when setting the values of the explanatory variables with indices from the set $J \cup L$ equal to their values in $\underline{x}_*$, and the expected prediction conditional on setting the values of the explanatory variables with indices from the set J equal to their values in $\underline{x}_*$.

In particular, for the l-th explanatory variable, let

$$\Delta^{l|J}(\underline{x}_*) \equiv \Delta^{\{l\}|J}(\underline{x}_*) = E_{\underline{X}}\{f(\underline{X})|X^{j_1} = x_*^{j_1}, \ldots, X^{j_K} = x_*^{j_K}, X^l = x_*^l\} -$$
$$E_{\underline{X}}\{f(\underline{X})|X^{j_1} = x_*^{j_1}, \ldots, X^{j_K} = x_*^{j_K}\}. \tag{6.10}$$

Thus, $\Delta^{l|J}$ is the change between the expected prediction, when setting the values of the explanatory variables with indices from the set $J \cup \{l\}$ equal to their values in $\underline{x}_*$, and the expected prediction conditional on setting the values of the explanatory variables with indices from the set J equal to their values in $\underline{x}_*$. Note that, if $J = \emptyset$, then

$$\Delta^{l|\emptyset}(\underline{x}_*) = E_{\underline{X}}\{f(\underline{X})|X^l = x_*^l\} - E_{\underline{X}}\{f(\underline{X})\} = E_{\underline{X}}\{f(\underline{X})|X^l = x_*^l\} - v_0. \tag{6.11}$$

It follows that

$$v(j, \underline{x}_*) = \Delta^{j|\{1, \ldots, j-1\}}(\underline{x}_*) = \Delta^{\{1, \ldots, j\}|\emptyset}(\underline{x}_*) - \Delta^{\{1, \ldots, j-1\}|\emptyset}(\underline{x}_*). \tag{6.12}$$

As it was mentioned in Section 6.2, for models that include interactions, the value of the variable-importance measure $v(j, \underline{x}_*)$ depends on the order of conditioning on explanatory variables. A heuristic approach to address this issue consists of choosing an order in which the variables with the largest contributions are selected first. In particular, the following two-step procedure can be considered. In the first step, the ordering is chosen based on the decreasing values of $|\Delta^{k|\emptyset}(\underline{x}_*)|$. Note that the use of absolute values is needed because the variable contributions can be positive or negative. In the second step, the variable-importance measure for the j-th variable is calculated as

$$v(j, \underline{x}_*) = \Delta^{j|J}(\underline{x}_*),$$

where

$$J = \{k : |\Delta^{k|\emptyset}(\underline{x}_*)| < |\Delta^{j|\emptyset}(\underline{x}_*)|\}.$$

That is, J is the set of indices of explanatory variables with scores $|\Delta^{k|\emptyset}(\underline{x}_*)|$ smaller than the corresponding score for variable j.

The time complexity of each of the two steps of the procedure is $O(p)$, where p is the number of explanatory variables.

Note, that there are also other possible approaches to the problem of calculation of variable attributions. One consists of identifying the interactions that cause a difference in variable-importance measures for different orderings and focusing on those interactions. This approach is discussed in Chapter 7. The other one consists of calculating an average value of the variance-importance measure across all possible orderings. This approach is presented in Chapter 8.

6.4   Example: Titanic data

Let us consider the random forest model `titanic_rf` (see Section 4.2.2) and passenger Johnny D (see Section 4.2.5) as the instance of interest in the Titanic data.

The mean of model's predictions for all passengers is equal to $v_0 = 0.2353095$. Table 6.1 presents the scores $|\Delta^{j|\emptyset}(\underline{x}_*)|$ and the expected values $E_X\{f(X)|X^j = x_*^j\}$. Note that, given (6.11) and the fact that $E_X\{f(X)|X^j = x_*^j\} > v_0$ for all variables, we have got $E_X\{f(X)|X^j = x_*^j\} = |\Delta^{j|\emptyset}(\underline{x}_*)| + v_0$.

TABLE 6.1: Expected values $E_X\{f(X)|X^j = x_*^j\}$ and scores $|\Delta^{j|\emptyset}(\underline{x}_*)|$ for the random forest model and Johnny D for the Titanic data. The scores are sorted in decreasing order.

variable j	$E_X\{f(X)|X^j = x_*^j\}$	$|\Delta^{j|\emptyset}(\underline{x}_*)|$
age = 8	0.5051210	0.2698115
class = 1st	0.4204449	0.1851354
fare = 72	0.3785383	0.1432288
gender = male	0.1102873	0.1250222
embarked = Southampton	0.2246035	0.0107060
sibsp = 0	0.2429597	0.0076502
parch = 0	0.2322655	0.0030440

Based on the ordering defined by the scores $|\Delta^{j|\emptyset}(\underline{x}_*)|$ from Table 6.1, we can compute the variable-importance measures based on the sequential contributions $\Delta^{j|J}(\underline{x}_*)$. The computed values are presented in Table 6.2.

TABLE 6.2: Variable-importance measures $\Delta^{j|J}(\underline{x}_*)$, with $J = \{1,\ldots,j\}$, for the random forest model and Johnny D for the Titanic data, computed by using the ordering of variables defined in Table 6.1.

variable j	$E_{\underline{X}}\left\{f(\underline{X})|\underline{X}^J = \underline{x}_*^J\right\}$	$\Delta^{j|J}(\underline{x}_*)$
intercept (v_0)	0.2353095	0.2353095
age = 8	0.5051210	0.2698115
class = 1st	0.5906969	0.0855759
fare = 72	0.5443561	-0.0463407
gender = male	0.4611518	-0.0832043
embarked = Southampton	0.4584422	-0.0027096
sibsp = 0	0.4523398	-0.0061024
parch = 0	0.4220000	-0.0303398
prediction	0.4220000	0.4220000

Results from Table 6.2 are presented in Figure 6.3. The plot indicates that the largest positive contributions to the predicted probability of survival for Johnny D come from explanatory variables *age* and *class*. The contributions of the remaining variables are smaller (in absolute values) and negative.

6.5 Pros and cons

BD plots offer a model-agnostic approach that can be applied to any predictive model that returns a single number for a single observation (instance). The approach offers several advantages. The plots are, in general, easy to understand. They are compact; results for many explanatory variables can be presented in a limited space. The approach reduces to an intuitive interpretation for linear models. Numerical complexity of the BD algorithm is linear in the number of explanatory variables.

An important issue is that BD plots may be misleading for models including interactions. This is because the plots show only the additive attributions. Thus, the choice of the ordering of the explanatory variables that is used in the calculation of the variable-importance measures is important. Also, for models with a large number of variables, BD plots may be complex and include many explanatory variables with small contributions to the instance prediction.

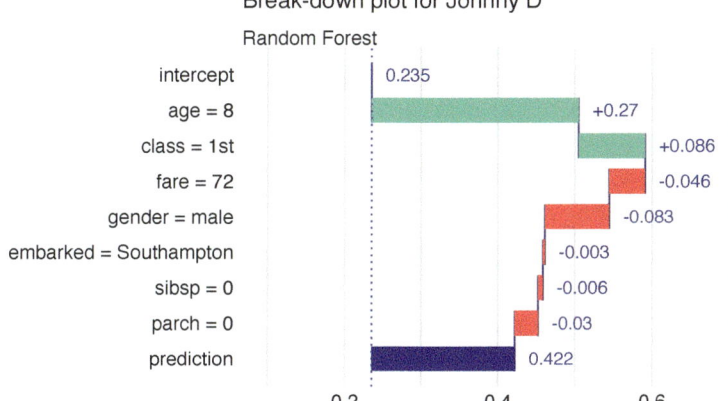

FIGURE 6.3 Break-down plot for the random forest model and Johnny D for the Titanic data.

To address the issue of the dependence of the variable-importance measure on the ordering of the explanatory variables, the heuristic approach described in Section 6.3.2 can be applied. Alternative approaches are described in Chapters 7 and 8.

6.6 Code snippets for R

In this section, we use the `DALEX` package, which is a wrapper for the `iBreakDown` R package (Gosiewska and Biecek, 2019). The package covers all methods presented in this chapter. It is available on `CRAN` and `GitHub`.

For illustration purposes, we use the `titanic_rf` random forest model for the Titanic data developed in Section 4.2.2. Recall that the model is developed to predict the probability of survival for passengers of Titanic. Instance-level explanations are calculated for Henry, a 47-year-old passenger that travelled in the 1st class (see Section 4.2.5).

We first retrieve the `titanic_rf` model-object and the data frame for Henry via the `archivist` hooks, as listed in Section 4.2.7. We also retrieve the version of the `titanic` data with imputed missing values.

```
titanic_imputed <- archivist::aread("pbiecek/models/27e5c")
titanic_rf <- archivist:: aread("pbiecek/models/4e0fc")
(henry <- archivist::aread("pbiecek/models/a6538"))


##    class gender age sibsp parch fare   embarked
## 1   1st    male  47     0     0   25  Cherbourg
```

Then we construct the explainer for the model by using function `explain()` from the `DALEX` package (see Section 4.2.6). We also load the `randomForest` package, as the model was fitted by using function `randomForest()` from this package (see Section 4.2.2) and it is important to have the corresponding `predict()` function available.

```
library("randomForest")
library("DALEX")
explain_rf <- DALEX::explain(model = titanic_rf,
                        data = titanic_imputed[, -9],
                           y = titanic_imputed$survived == "yes",
                       label = "Random Forest")
```

The explainer object allows uniform access to predictive models regardless of their internal structure. With this object, we can proceed to the model analysis.

6.6.1   Basic use of the `predict_parts()` function

The `DALEX::predict_parts()` function decomposes model predictions into parts that can be attributed to individual variables. It calculates the variable-attribution measures for a selected model and an instance of interest. The object obtained as a result of applying the function is a data frame containing the calculated measures.

In the simplest call, the function requires three arguments:

- `explainer` - an explainer-object, created with function `DALEX::explain()`;
- `new_observation` - an observation to be explained; it should be a data frame with a structure that matches the structure of the dataset used for fitting of the model;
- `type` - the method for calculation of variable attribution; the possible methods are `"break_down"` (the default), `"shap"`, `"oscillations"`, and `"break_down_interactions"`.

In the code below, the argument `type = "break_down"` is explicitly used. The code essentially provides the variable-importance values $\Delta^{j|\{1,\ldots,j\}}(\underline{x}_*)$.

```
bd_rf <- predict_parts(explainer = explain_rf,
                new_observation = henry,
                           type = "break_down")
bd_rf
```

```
##                                            contribution
## Random Forest: intercept                         0.235
## Random Forest: class = 1st                       0.185
## Random Forest: gender = male                    -0.124
## Random Forest: embarked = Cherbourg              0.105
## Random Forest: age = 47                         -0.092
```

```
## Random Forest: fare = 25                    -0.030
## Random Forest: sibsp = 0                    -0.032
## Random Forest: parch = 0                    -0.001
## Random Forest: prediction                    0.246
```

By applying the generic `plot()` function to the object resulting from the application of the `predict_parts()` function we obtain a BD plot.

```
plot(bd_rf)
```

The resulting plot is shown in Figure 6.4. It can be used to compare the explanatory-variable attributions obtained for Henry with those computed for Johnny D (see Figure 6.3). Both explanations refer to the same random forest model. We can see that the predicted survival probability for Henry (0.246) is almost the same as the mean prediction (0.235), while the probability for Johnny D is higher (0.422). For Johnny D, this result can be mainly attributed to the positive contribution of *age* and *class*. For Henry, *class* still contributes positively to the chances of survival, but the effect of *age* is negative. For both passengers the effect of *gender* is negative. Thus, one could conclude that the difference in the predicted survival probabilities is mainly due to the difference in the age of Henry and Johnny D.

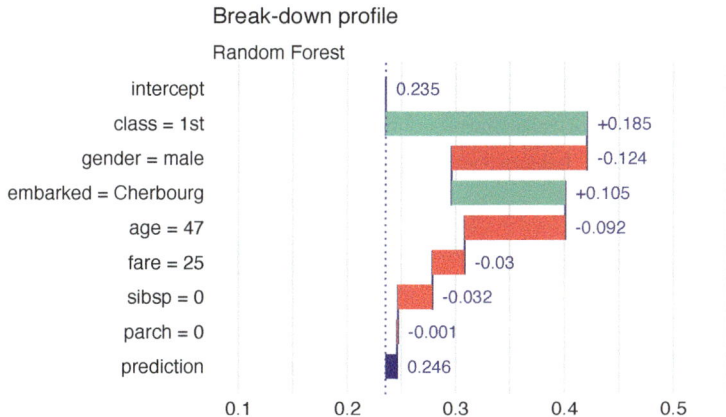

FIGURE 6.4 Break-down plot for the random forest model and Henry for the Titanic data, obtained by the generic `plot()` function in R.

6.6.2 Advanced use of the `predict_parts()` function

Apart from the `explainer`, `new_observation`, and `type` arguments, function `predict_parts()` allows additional ones. The most commonly used are:

- `order` - a vector of characters (column names) or integers (column indexes) that specify the order of explanatory variables to be used for computing the variable-importance measures; if not specified (default), then a one-step heuristic is used to determine the order;

- `keep_distributions` - a logical value (`FALSE` by default); if `TRUE`, then additional diagnostic information about conditional distributions of predictions is stored in the resulting object and can be plotted with the generic `plot()` function.

In what follows, we illustrate the use of the arguments.

First, we specify the ordering of the explanatory variables. Toward this end, we can use integer indexes or variable names. The latter option is preferable in most cases because of transparency. Additionally, to reduce clutter in the plot, we set `max_features = 3` argument in the `plot()` function.

```
bd_rf_order <- predict_parts(explainer = explain_rf,
                    new_observation = henry,
                             type = "break_down",
                order = c("class", "age", "gender", "fare",
                         "parch", "sibsp", "embarked"))
plot(bd_rf_order, max_features = 3)
```

The resulting plot is presented in Figure 6.5. It is worth noting that the attributions for variables *gender* and *fare* do differ from those shown in Figure 6.4. This is the result of the change of the ordering of variables used in the computation of the attributions.

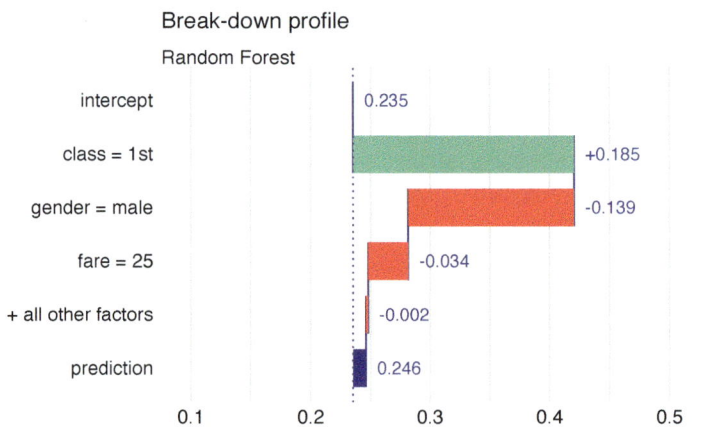

FIGURE 6.5 Break-down plot for the top three variables for the random forest model and Henry for the Titanic data.

We can use the `keep_distributions = TRUE` argument to enrich the resulting object with additional information about conditional distributions of predicted values. Subsequently, we can apply the `plot_distributions = TRUE` argument in the `plot()` function to present the distributions as violin plots.

```
bd_rf_distr <- predict_parts(explainer = explain_rf,
                  new_observation = henry,
                              type = "break_down",
         order = c("age", "class", "fare", "gender",
                   "embarked", "sibsp", "parch"),
              keep_distributions = TRUE)
plot(bd_rf_distr, plot_distributions = TRUE)
```

The resulting plot is presented in Figure 6.6. Red dots indicate the mean model's predictions. Thin grey lines between violin plots indicate changes in predictions for individual observations. They can be used to track how the model's predictions change after consecutive conditionings. A similar code was used to create the plot in panel A of Figure 6.1 for Johnny D.

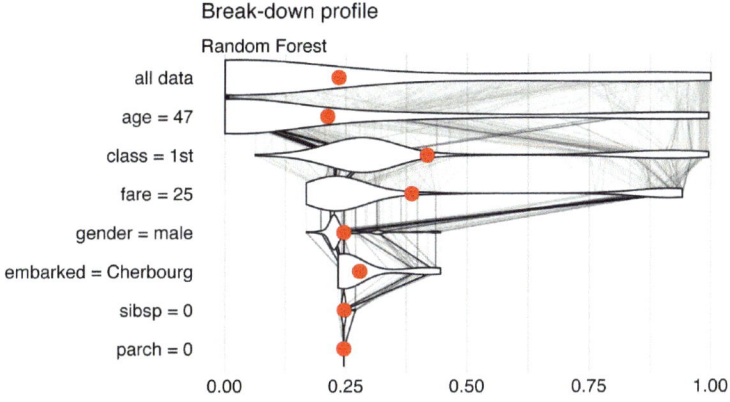

FIGURE 6.6 Break-down plot with violin plots summarizing distributions of predicted values for a selected order of explanatory variables for the random forest model and Henry for the Titanic data.

6.7 Code snippets for Python

In this section, we use the **dalex** library in Python. The package covers all methods presented in this chapter. It is available on **pip** and **GitHub**.

For illustration purposes, we use the **titanic_rf** random forest model for the Titanic data developed in Section 4.3.2. Recall that the model is developed to predict the probability of survival for passengers of Titanic. Instance-level explanations are calculated for Henry, a 47-year-old passenger that travelled in the first class (see Section 4.3.5).

In the first step, we create an explainer object that will provide a uniform interface for the predictive model. We use the `Explainer()` constructor for this purpose (see Section 4.3.6).

```
import pandas as pd
henry = pd.DataFrame({'gender'   : ['male'],
                      'age'       : [47],
                      'class'     : ['1st'],
                      'embarked'  : ['Cherbourg'],
                      'fare'      : [25],
                      'sibsp'     : [0],
                      'parch'     : [0]},
                      index = ['Henry'])


import dalex as dx
titanic_rf_exp = dx.Explainer(titanic_rf, X, y,
                label = "Titanic RF Pipeline")
```

To apply the break-down method we use the `predict_parts()` method. The first argument indicates the data for the observation for which the attributions are to be calculated. The `type` argument specifies the method of calculation of attributions. Results are stored in the `result` field.

```
bd_henry = titanic_rf_exp.predict_parts(henry,
           type = 'break_down')
bd_henry.result
```

	variable_name	variable_value	variable	cumulative	contribution	sign	position	label
0	intercept	1	intercept	0.322108	0.322108	1.0	8	Titanic RF Pipeline
1	class	1st	class = 1st	0.387939	0.065832	1.0	7	Titanic RF Pipeline
2	embarked	Cherbourg	embarked = Cherbourg	0.416226	0.028287	1.0	6	Titanic RF Pipeline
3	fare	25.0	fare = 25.0	0.427836	0.011609	1.0	5	Titanic RF Pipeline
4	sibsp	0.0	sibsp = 0.0	0.427544	-0.000291	-1.0	4	Titanic RF Pipeline
5	parch	0.0	parch = 0.0	0.423762	-0.003782	-1.0	3	Titanic RF Pipeline
6	age	47.0	age = 47.0	0.414922	-0.008840	-1.0	2	Titanic RF Pipeline
7	gender	male	gender = male	0.300822	-0.114100	-1.0	1	Titanic RF Pipeline
8			prediction	0.300822	0.300822	1.0	0	Titanic RF Pipeline

To obtain a waterfall chart we can use the `plot()` method. It generates an interactive chart based on the `plotly` library.

```
bd_henry.plot()
```

The resulting plot is presented in Figure 6.7.

Advanced users can make use of the `order` argument of the `predict_parts()` method. It allows forcing a specific order of variables in the break-down method. Also, if the model includes many explanatory variables, the waterfall chart

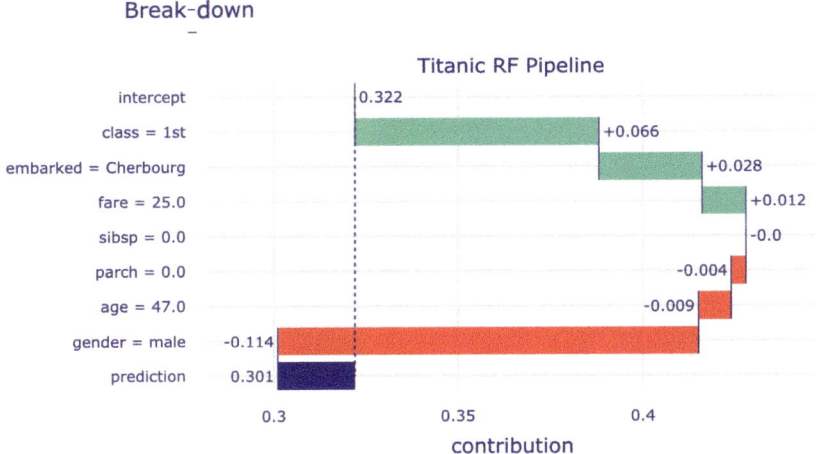

FIGURE 6.7 Break-down plot for the random forest model and Henry for the Titanic data, obtained by the `plot()` method in Python.

may be hard to read. In this situation, the `max_vars` argument can be used in the `plot()` method to limit the number of variables presented in the graph.

```python
import numpy as np

bd_henry = titanic_rf_exp.predict_parts(henry,
        type = 'break_down',
        order = np.array(['gender', 'class', 'age',
            'embarked', 'fare', 'sibsp', 'parch']))
bd_henry.plot(max_vars = 5)
```

The resulting plot is presented in Figure 6.8.

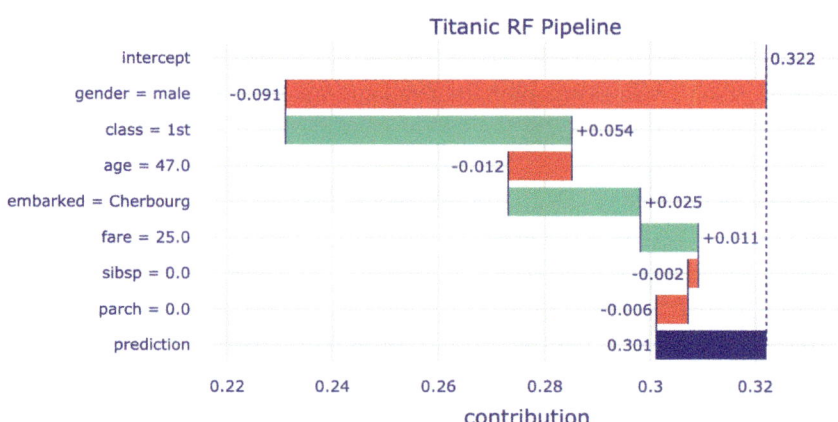

FIGURE 6.8 Break-down plot for a limited number of explanatory variables in a specified order for the random forest model and Henry for the Titanic data, obtained by the `plot()` method in Python.

7

Break-down Plots for Interactions

In Chapter 6, we presented a model-agnostic approach to the calculation of the attribution of an explanatory variable to a model's predictions. However, for some models, like models with interactions, the results of the method introduced in Chapter 6 depend on the ordering of the explanatory variables that are used in computations.

In this chapter, we present an algorithm that addresses the issue. In particular, the algorithm identifies interactions between pairs of variables and takes them into account when constructing break-down (BD) plots. In our presentation, we focus on pairwise interactions that involve pairs of explanatory variables, but the algorithm can be easily extended to interactions involving a larger number of variables.

7.1 Intuition

Interaction (deviation from additivity) means that the effect of an explanatory variable depends on the value(s) of other variable(s). To illustrate such a situation, we use the Titanic dataset (see Section 4.1). For the sake of simplicity, we consider only two variables, *age* and *class*. *Age* is a continuous variable, but we will use a dichotomized version of it, with two levels: boys (0-16 years old) and adults (17+ years old). Also, for *class*, we will consider just "2nd class" and "other".

Table 7.1 shows percentages of survivors for boys and adult men travelling in the second class and other classes on Titanic. Overall, the proportion of survivors among males is 20.5%. However, among boys in the second class, the proportion is 91.7%. How do *age* and *class* contribute to this higher survival probability? Let us consider the following two explanations.

Explanation 1:

The overall probability of survival for males is 20.5%, but for the male passengers from the second class, the probability is even lower, i.e., 13.5%. Thus, the effect of the travel class is negative, as it decreases the probability of survival by 7 percentage points. Now, if, for male passengers of the second class, we

consider their age, we see that the survival probability for boys increases by
78.2 percentage points, from 13.5% (for a male in the second class) to 91.7%.
Thus, by considering first the effect of *class*, and then the effect of *age*, we can
conclude the effect of -7 percentage points for *class* and $+78.2$ percentage
points for *age* (being a boy).

Explanation 2:

The overall probability of survival for males is 20.5%, but for boys the probability is higher, i.e., 40.7%. Thus, the effect of *age* (being a boy) is positive, as
it increases the survival probability by 20.2 percentage points. On the other
hand, for boys, travelling in the second class increases the probability further
from 40.7% overall to 91.7%. Thus, by considering first the effect of *age*, and
then the effect of *class*, we can conclude the effect of $+20.2$ percentage points
for *age* (being a boy) and $+51$ percentage points for *class*.

TABLE 7.1: Proportion of survivors for men on Titanic.

Class	Boys (0-16)	Adults (>16)	Total
2nd	$11/12 = 91.7\%$	$13/166 = 7.8\%$	$24/178 = 13.5\%$
other	$22/69 = 31.9\%$	$306/1469 = 20.8\%$	$328/1538 = 21.3\%$
Total	$33/81 = 40.7\%$	$319/1635 = 19.5\%$	$352/1716 = 20.5\%$

Thus, by considering the effects of *class* and *age* in a different order, we get
very different attributions (contributions attributed to the variables). This is
because there is an interaction: the effect of *class* depends on *age* and *vice
versa*. In particular, from Table 7.1 we could conclude that the overall effect
of the second class is negative (-7 percentage points), as it decreases the
probability of survival from 20.5% to 13.5%. On the other hand, the overall
effect of being a boy is positive ($+20.2$ percentage points), as it increases
the probability of survival from 20.5% to 40.7%. Based on those effects, we
would expect a probability of $20.5\% - 7\% + 20.2\% = 33.7\%$ for a boy in the
second class. However, the observed proportion of survivors is much higher,
91.7%. The difference $91.7\% - 33.7\% = 58\%$ is the interaction effect. We can
interpret it as an additional effect of the second class specific for boys, or as
an additional effect of being a boy for the male passengers travelling in the
second class.

The example illustrates that interactions complicate the evaluation of the
importance of explanatory variables with respect to a model's predictions. In
the next section, we present an algorithm that allows including interactions in
the BD plots.

7.2 Method

Identification of interactions in the model is performed in three steps (Gosiewska and Biecek, 2019):

1. For each explanatory variable, compute $\Delta^{j|\emptyset}(\underline{x}_*)$ as in equation (6.11) in Section 6.3.2. The measure quantifies the additive contribution of each variable to the instance prediction.
2. For each pair of explanatory variables, compute $\Delta^{\{i,j\}|\emptyset}(\underline{x}_*)$ as in equation (6.11) in Section 6.3.2, and then the "net effect" of the interaction

$$\Delta_I^{\{i,j\}}(\underline{x}_*) \equiv \Delta^{\{i,j\}|\emptyset}(\underline{x}_*) - \Delta^{i|\emptyset}(\underline{x}_*) - \Delta^{j|\emptyset}(\underline{x}_*). \qquad (7.1)$$

   Note that $\Delta^{\{i,j\}|\emptyset}(\underline{x}_*)$ quantifies the joint contribution of a pair of variables. Thus, $\Delta_I^{\{i,j\}}(\underline{x}_*)$ measures the contribution related to the deviation from additivity, i.e., to the interaction between the i-th and j-th variable.
3. Rank the so-obtained measures for individual explanatory variables and interactions to determine the final ordering for computing the variable-importance measures. Using the ordering, compute variable-importance measures $v(j, \underline{x}_*)$, as defined in equation (6.12) in Section 6.3.2.

The time complexity of the first step is $O(p)$, where p is the number of explanatory variables. For the second step, the complexity is $O(p^2)$, while for the third step it is $O(p)$. Thus, the time complexity of the entire procedure is $O(p^2)$.

7.3 Example: Titanic data

Let us consider the random forest model `titanic_rf` (see Section 4.2.2) and passenger Johnny D (see Section 4.2.5) as the instance of interest in the Titanic data.

Table 7.2 presents single-variable contributions $\Delta^{j|\emptyset}(\underline{x}_*)$, paired-variable contributions $\Delta^{\{i,j\}|\emptyset}(\underline{x}_*)$, and interaction contributions $\Delta_I^{\{i,j\}}(\underline{x}_*)$ for each explanatory variable and each pair of variables. All the measures are calculated for Johnny D, the instance of interest.

TABLE 7.2: Paired-variable contributions $\Delta^{\{i,j\}|\emptyset}(\underline{x}_*)$, interaction contributions $\Delta_I^{\{i,j\}}(\underline{x}_*)$, and single-variable contributions $\Delta^{j|\emptyset}(\underline{x}_*)$ for the random forest model and Johnny D for the Titanic data.

Variable	$\Delta^{\{i,j\}|\emptyset}(\underline{x}_*)$	$\Delta_I^{\{i,j\}}(\underline{x}_*)$	$\Delta^{i|\emptyset}(\underline{x}_*)$
age			0.270
fare:class	0.098	-0.231	
class			0.185
fare:age	0.249	-0.164	
fare			0.143
gender			-0.125
age:class	0.355	-0.100	
age:gender	0.215	0.070	
fare:gender			
embarked			-0.011
embarked:age	0.269	0.010	
parch:gender	-0.136	-0.008	
sibsp			0.008
sibsp:age	0.284	0.007	
sibsp:class	0.187	-0.006	
embarked:fare	0.138	0.006	
sibsp:gender	-0.123	-0.005	
fare:parch	0.145	0.005	
parch:sibsp	0.001	-0.004	
parch			-0.003
parch:age	0.264	-0.002	
embarked:gender	-0.134	0.002	
embarked:parch	-0.012	0.001	
fare:sibsp	0.152	0.001	
embarked:class	0.173	-0.001	
gender:class	0.061	0.001	
embarked:sibsp	-0.002	0.001	
parch:class	0.183	0.000	

The table illustrates the calculation of the contributions of interactions. For instance, the additive contribution of *age* is equal to 0.270, while for *fare* it is equal to 0.143. The joint contribution of these two variables is equal to 0.249. Hence, the contribution attributed to the interaction is equal to $0.249 - 0.270 - 0.143 = -0.164$.

Note that the rows of Table 7.2 are sorted according to the absolute value of the net contribution of the single explanatory variable or the net contribution of the interaction between two variables. For a single variable, the net contribution is simply measured by $\Delta^{j|\emptyset}(\underline{x}_*)$, while for an interaction it is given by $\Delta_I^{\{i,j\}}(\underline{x}_*)$.

In this way, if two variables are important and there is little interaction, then the net contribution of the interaction is smaller than the contribution of any of the two variables. Consequently, the interaction will be ranked lower. This is the case, for example, of variables *age* and *gender* in Table 7.2. On the other hand, if the interaction is important, then its net contribution will be larger than the contribution of any of the two variables. This is the case, for example, of variables *fare* and *class* in Table 7.2.

Based on the ordering of the rows in Table 7.2, the following sequence of variables is identified as informative:

- *age*, because it has the largest (in absolute value) net contribution equal to 0.270;
- *fare:class* interaction, because its net contribution (-0.231) is the second largest (in absolute value);
- *gender*, because variables *class* and *fare* are already accounted for in the *fare:class* interaction and the net contribution of *gender*, equal to 0.125, is the largest (in absolute value) among the remaining variables and interactions;
- *embarked* harbor (based on a similar reasoning as for *gender*);
- then *sibsp* and *parch* as variables with the smallest net contributions (among single variables), which are larger than the contribution of their interaction.

Table 7.3 presents the variable-importance measures computed by using the following ordering of explanatory variables and their pairwise interactions: *age*, *fare:class*, *gender*, *embarked*, *sibsp*, and *parch*. The table presents also the conditional expected values (see equations (6.6) and (6.12) in Section 6.3.2)

$$E_{\underline{X}}\left\{f(\underline{X})|\underline{X}^{\{1,\ldots,j\}} = \underline{x}_*^{\{1,\ldots,j\}}\right\} = v_0 + \sum_{k=1}^{j} v(k, \underline{x}_*) = v_0 + \Delta^{\{1,\cdots j\}|\emptyset}(\underline{x}_*).$$

Note that the expected value presented in the last row, 0.422, corresponds to the model's prediction for the instance of interest, passenger Johnny D.

> TABLE 7.3: Variable-importance measures $v(j, \underline{x}_*)$ and the conditonal expected values $v_0 + \sum_{k=1}^{j} v(k, \underline{x}_*)$ computed by using the sequence of variables *age*, *fare:class*, *gender*, *embarked*, *sibsp*, and *parch* for the random forest model and Johnny D for the Titanic data.

Variable	j	$v(j, \underline{x}_*)$	$v_0 + \sum_{k=1}^{j} v(k, \underline{x}_*)$
intercept (v_0)			0.235
age = 8	1	0.269	0.505
fare:class = 72:1st	2	0.039	0.544
gender = male	3	-0.083	0.461
embarked = Southampton	4	-0.002	0.458
sibsp = 0	5	-0.006	0.452
parch = 0	6	-0.030	0.422

Figure 7.1 presents the interaction-break-down (iBD) plot corresponding to the results shown in Table 7.3. The interaction between *fare* and *class* variables is included in the plot as a single bar. As the effects of these two variables cannot be disentangled, the plot uses just that single bar to represent the contribution of both variables. Table 7.2 indicates that *class* alone would increase the mean prediction by 0.185, while *fare* would increase the mean prediction by 0.143. However, taken together, they increase the average prediction only by 0.098. A possible explanation of this negative interaction could be that, while the ticket fare of 72 is high on average, it is actually below the median when the first-class passengers are considered. Thus, if first-class passengers with "cheaper" tickets, as Johnny D, were, for instance, placed in cabins that made it more difficult to reach a lifeboat, this could lead to lower chances of survival as compared to other passengers from the same class (though the chances could be still higher as compared to passengers from other, lower travel classes).

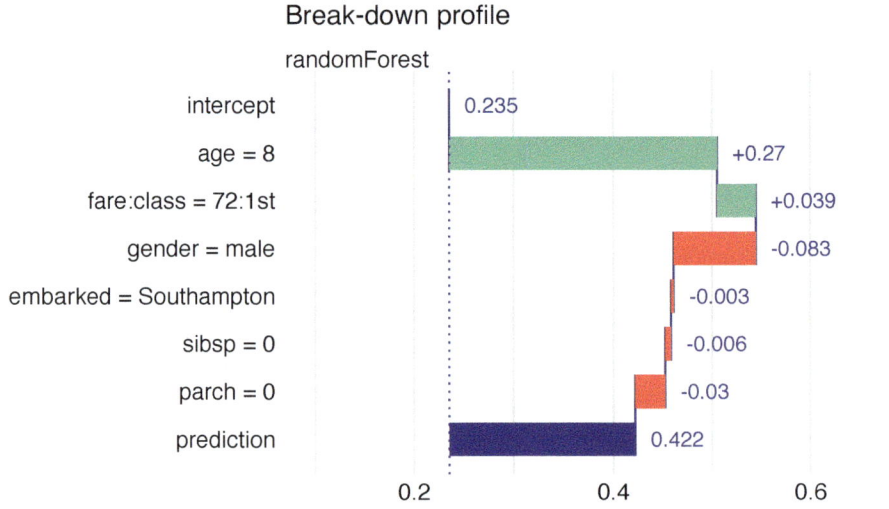

FIGURE 7.1 Break-down plot with interactions for the random forest model and Johnny D for the Titanic data.

7.4 Pros and cons

iBD plots share many advantages and disadvantages of BD plots for models without interactions (see Section 6.5). However, in the case of models with interactions, iBD plots provide more correct explanations.

Though the numerical complexity of the iBD procedure is quadratic, it may be time-consuming in case of models with a large number of explanatory variables.

For a model with p explanatory variables, we have got to calculate $p*(p+1)/2$ net contributions for single variables and pairs of variables. For datasets with a small number of observations, the calculations of the net contributions will be subject to a larger variability and, therefore, larger randomness in the ranking of the contributions.

It is also worth noting that the presented procedure of identification of interactions is not based on any formal statistical-significance test. Thus, the procedure may lead to false-positive findings and, especially for small sample sizes, false-negative errors.

7.5 Code snippets for R

In this section, we use the `DALEX` package, which is a wrapper for `iBreakDown` R package. The package covers all methods presented in this chapter. It is available on `CRAN` and `GitHub`.

For illustration purposes, we use the `titanic_rf` random forest model for the Titanic data developed in Section 4.2.2. Recall that the model is constructed to predict the probability of survival for passengers of Titanic. Instance-level explanations are calculated for Henry, a 47-year-old passenger that travelled in the first class (see Section 4.2.5).

First, we retrieve the `titanic_rf` model-object and the data for Henry via the `archivist` hooks, as listed in Section 4.2.7. We also retrieve the version of the `titanic` data with imputed missing values.

```r
titanic_imputed <- archivist::aread("pbiecek/models/27e5c")
titanic_rf <- archivist:: aread("pbiecek/models/4e0fc")
(henry <- archivist::aread("pbiecek/models/a6538"))
```

```
  class gender age sibsp parch fare   embarked
1   1st   male  47     0     0   25 Cherbourg
```

Then we construct the explainer for the model by using the function `explain()` from the `DALEX` package (see Section 4.2.6). Note that, beforehand, we have got to load the `randomForest` package, as the model was fitted by using function `randomForest()` from this package (see Section 4.2.2) and it is important to have the corresponding `predict()` function available.

```r
library("DALEX")
library("randomForest")
explain_rf <- DALEX::explain(model = titanic_rf,
                      data = titanic_imputed[, -9],
                         y = titanic_imputed$survived == "yes",
                     label = "Random Forest")
```

The key function to construct iBD plots is the `DALEX::predict_parts()` function. The use of the function has already been explained in Section 6.6. In order to perform calculations that allow obtaining iBD plots, the required argument is `type = "break_down_interactions"`.

```
bd_rf <- predict_parts(explainer = explain_rf,
                   new_observation = henry,
                           type = "break_down_interactions")
bd_rf
```

```
##                                               contribution
## Random Forest: intercept                             0.235
## Random Forest: class = 1st                           0.185
## Random Forest: gender = male                        -0.124
## Random Forest: embarked:fare = Cherbourg:25          0.107
## Random Forest: age = 47                             -0.125
## Random Forest: sibsp = 0                            -0.032
## Random Forest: parch = 0                            -0.001
## Random Forest: prediction                            0.246
```

We can compare the obtained variable-importance measures to those reported for Johnny D in Table 7.3. For Henry, the most important positive contribution comes from *class*, while for Johnny D it is *age*. Interestingly, for Henry, a positive contribution of the interaction between *embarked* harbour and *fare* is found. For Johnny D, a different interaction was identified: for *fare* and *class*. Finding an explanation for this difference is not straightforward. In any case, in those two instances, the contribution of fare appears to be modified by effects of other variable(s), i.e., its effect is not purely additive.

By applying the generic `plot()` function to the object created by the `DALEX::predict_parts()` function we obtain the iBD plot.

```
plot(bd_rf)
```

The resulting iBD plot for Henry is shown in Figure 7.2. It can be compared to the iBD plot for Johnny D that is presented in Figure 7.1.

7.6   Code snippets for Python

In this section, we use the `dalex` library for Python. The package covers all methods presented in this chapter. It is available on `pip` and `GitHub`.

For illustration purposes, we use the `titanic_rf` random forest model for the Titanic data developed in Section 4.3.2. Instance-level explanations are calculated for Henry, a 47-year-old passenger that travelled in the first class (see Section 4.3.5).

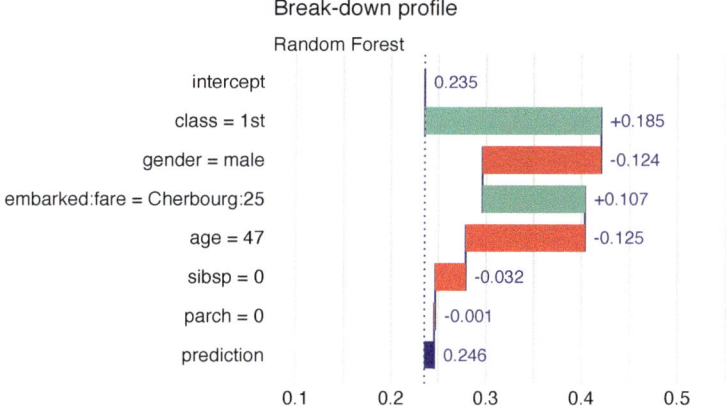

FIGURE 7.2 Break-down plot with interactions for the random forest model and Henry for the Titanic data, obtained by applying the generic `plot()` function in R.

In the first step, we create an explainer-object that provides a uniform interface for the predictive model. We use the `Explainer()` constructor for this purpose (see Section 4.3.6).

```
import pandas as pd
henry = pd.DataFrame({'gender': ['male'], 'age': [47],
        'class': ['1st'],
        'embarked': ['Cherbourg'], 'fare': [25],
        'sibsp': [0], 'parch': [0]},
        index = ['Henry'])
import dalex as dx
titanic_rf_exp = dx.Explainer(titanic_rf, X, y,
        label = "Titanic RF Pipeline")
```

To calculate the attributions with the break-down method with interactions, we use the `predict_parts()` method with `type='break_down_interactions'` argument (see Section 6.7). The first argument indicates the data for the observation for which the attributions are to be calculated.

Interactions are often weak and their net effects are not larger than the contributions of individual variables. If we would like to increase our preference for interactions, we can use the `interaction_preference` argument. The default value of 1 means no preference, while larger values indicate a larger preference. Results are stored in the `result` field.

```
bd_henry = titanic_rf_exp.predict_parts(henry,
            type = 'break_down_interactions',
            interaction_preference = 10)
bd_henry.result
```

	variable_name	variable_value	variable	cumulative	contribution	sign	position	label
0	intercept	1	intercept	0.322108	0.322108	1.0	5	Titanic RF Pipeline
1	class:gender	1st:male	class:gender = 1st:male	0.285465	-0.036643	-1.0	4	Titanic RF Pipeline
2	fare:embarked	25.0:Cherbourg	fare:embarked = 25.0:Cherbourg	0.322390	0.036925	1.0	3	Titanic RF Pipeline
3	parch:sibsp	0.0:0.0	parch:sibsp = 0.0:0.0	0.314178	-0.008212	-1.0	2	Titanic RF Pipeline
4	age	47.0	age = 47.0	0.300822	-0.013357	-1.0	1	Titanic RF Pipeline
5			prediction	0.300822	0.300822	1.0	0	Titanic RF Pipeline

By applying the `plot()` method to the resulting object, we construct the corresponding iBD plot.

```
bd_henry.plot()
```

The resulting plot for Henry is shown in Figure 7.3.

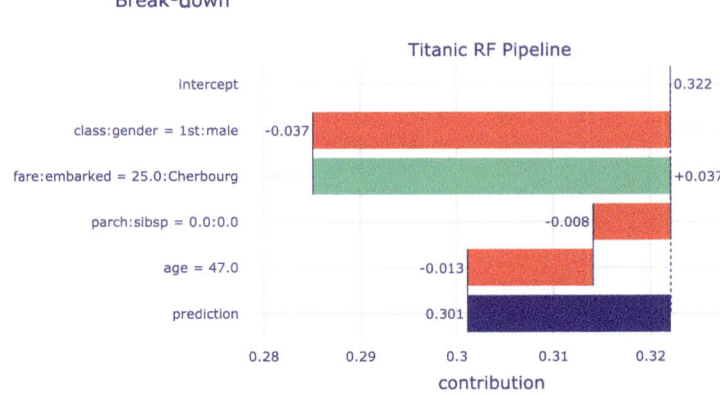

FIGURE 7.3 Break-down plot with interactions for the random forest model and Henry for the Titanic data, obtained by applying the `plot()` method in Python.

8

Shapley Additive Explanations (SHAP) for Average Attributions

---

In Chapter 6, we introduced break-down (BD) plots, a procedure for calculation of attribution of an explanatory variable for a model's prediction. We also indicated that, in the presence of interactions, the computed value of the attribution depends on the order of explanatory covariates that are used in calculations. One solution to the problem, presented in Chapter 6, is to find an ordering in which the most important variables are placed at the beginning. Another solution, described in Chapter 7, is to identify interactions and explicitly present their contributions to the predictions.

In this chapter, we introduce yet another approach to address the ordering issue. It is based on the idea of averaging the value of a variable's attribution over all (or a large number of) possible orderings. The idea is closely linked to "Shapley values" developed originally for cooperative games (Shapley, 1953). The approach was first translated to the machine-learning domain by Štrumbelj and Kononenko (2010) and Štrumbelj and Kononenko (2014). It has been widely adopted after the publication of the paper by Lundberg and Lee (2017) and Python's library for SHapley Additive exPlanations, SHAP (Lundberg, 2019). The authors of SHAP introduced an efficient algorithm for tree-based models (Lundberg et al., 2018). They also showed that Shapley values could be presented as a unification of a collection of different commonly used techniques for model explanations (Lundberg and Lee, 2017).

8.1 Intuition

Figure 8.1 presents BD plots for 10 random orderings (indicated by the order of the rows in each plot) of explanatory variables for the prediction for Johnny D (see Section 4.2.5) for the random forest model `titanic_rf` (see Section 4.2.2) for the Titanic dataset. The plots show clear differences in the contributions of various variables for different orderings. The most remarkable differences can be observed for variables *fare* and *class*, with contributions changing the sign depending on the ordering.

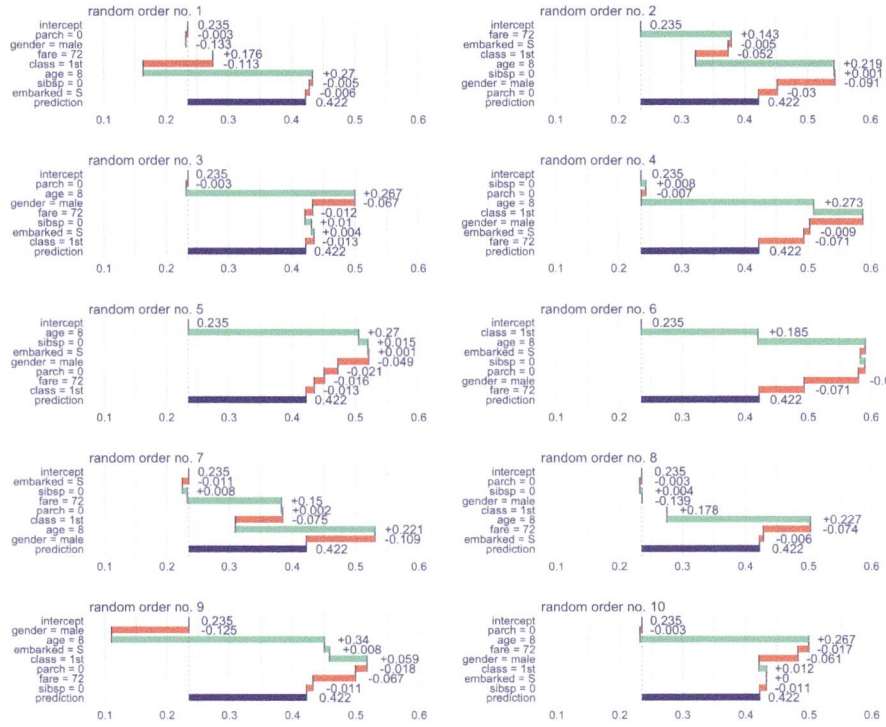

FIGURE 8.1 Break-down plots for 10 random orderings of explanatory variables for the prediction for Johnny D for the random forest model `titanic_rf` for the Titanic dataset. Each panel presents a single ordering, indicated by the order of the rows in the plot.

To remove the influence of the ordering of the variables, we can compute the mean value of the attributions. Figure 8.2 presents the averages, calculated over the ten orderings presented in Figure 8.1. Red and green bars present, respectively, the negative and positive averages Violet box plots summarize the distribution of the attributions for each explanatory variable across the different orderings. The plot indicates that the most important variables, from the point of view of the prediction for Johnny D, are *age*, *class*, and *gender*.

8.2   Method

SHapley Additive exPlanations (SHAP) are based on "Shapley values" developed by Shapley (1953) in the cooperative game theory. Note that the

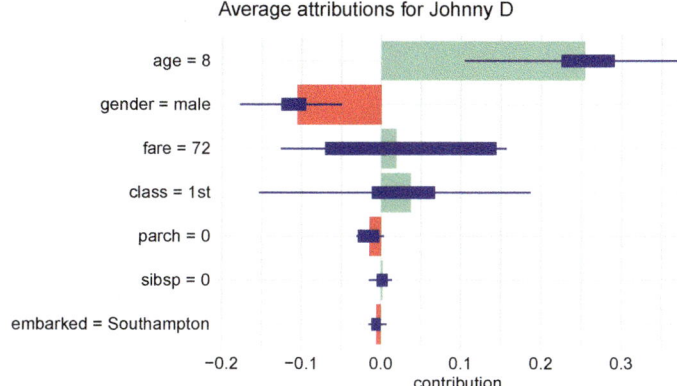

FIGURE 8.2 Average attributions for 10 random orderings. Red and green bars present the means. Box plots summarize the distribution of contributions for each explanatory variable across the orderings.

terminology may be confusing at first glance. Shapley values are introduced for cooperative games. SHAP is an acronym for a method designed for predictive models. To avoid confusion, we will use the term "Shapley values".

Shapley values are a solution to the following problem. A coalition of players cooperates and obtains a certain overall gain from the cooperation. Players are not identical, and different players may have different importance. Cooperation is beneficial, because it may bring more benefit than individual actions. The problem to solve is how to distribute the generated surplus among the players. Shapley values offer one possible fair answer to this question (Shapley, 1953).

Let's translate this problem to the context of a model's predictions. Explanatory variables are the players, while model $f()$ plays the role of the coalition. The payoff from the coalition is the model's prediction. The problem to solve is how to distribute the model's prediction across particular variables?

The idea of using Shapley values for evaluation of local variable-importance was introduced by Štrumbelj and Kononenko (2010). We will define the values using the notation introduced in Section 6.3.2.

Let us consider a permutation J of the set of indices $\{1, 2, \ldots, p\}$ corresponding to an ordering of p explanatory variables included in the model $f()$. Denote by $\pi(J, j)$ the set of the indices of the variables that are positioned in J before the j-th variable. Note that, if the j-th variable is placed as the first, then $\pi(J, j) = \emptyset$. Consider the model's prediction $f(\underline{x}_*)$ for a particular instance of interest $\underline{x}_*$. The Shapley value is defined as follows:

$$\varphi(\underline{x}_*, j) = \frac{1}{p!} \sum_J \Delta^{j|\pi(J,j)}(\underline{x}_*),\tag{8.1}$$

where the sum is taken over all $p!$ possible permutations (orderings of explanatory variables) and the variable-importance measure $\Delta^{j|J}(\underline{x}_*)$ was defined in equation (6.10) in Section 6.3.2. Essentially, $\varphi(\underline{x}_*, j)$ is the average of the variable-importance measures across all possible orderings of explanatory variables.

It is worth noting that the value of $\Delta^{j|\pi(J,j)}(\underline{x}_*)$ is constant for all permutations J that share the same subset $\pi(J, j)$. It follows that equation (8.1) can be expressed in an alternative form:

$$\begin{aligned}
\varphi(\underline{x}_*, j) &= \frac{1}{p!} \sum_{s=0}^{p-1} \sum_{\substack{S \subseteq \{1,\ldots,p\}\backslash\{j\} \\ |S|=s}} \left\{ s!(p-1-s)! \Delta^{j|S}(\underline{x}_*) \right\} \\
&= \frac{1}{p} \sum_{s=0}^{p-1} \sum_{\substack{S \subseteq \{1,\ldots,p\}\backslash\{j\} \\ |S|=s}} \left\{ \binom{p-1}{s}^{-1} \Delta^{j|S}(\underline{x}_*) \right\},
\end{aligned}\tag{8.2}$$

where $|S|$ denotes the cardinal number (size) of set S and the second sum is taken over all subsets S of explanatory variables, excluding the j-th one, of size s.

Note that the number of all subsets of sizes from 0 to $p-1$ is $2^p - 1$, i.e., it is much smaller than number of all permutations $p!$. Nevertheless, for a large p, it may be feasible to compute Shapley values neither using (8.1) nor (8.2). In that case, an estimate based on a sample of permutations may be considered. A Monte Carlo estimator was introduced by Štrumbelj and Kononenko (2014). An efficient implementation of computations of Shapley values for tree-based models was used in package SHAP (Lundberg and Lee, 2017).

From the properties of Shapley values for cooperative games it follows that, in the context of predictive models, they enjoy the following properties:

- *Symmetry*: if two explanatory variables j and k are interchangeable, i.e., if, for any set of explanatory variables $S \subseteq \{1,\ldots,p\} \setminus \{j, k\}$ we have got

$$\Delta^{j|S}(\underline{x}_*) = \Delta^{k|S}(\underline{x}_*),$$

then their Shapley values are equal:

$$\varphi(\underline{x}_*, j) = \varphi(\underline{x}_*, k).$$

- *Dummy feature*: if an explanatory variable j does not contribute to any prediction for any set of explanatory variables $S \subseteq \{1, \ldots, p\} \setminus \{j\}$, that is, if

$$\Delta^{j|S}(\underline{x}_*) = 0,$$

then its Shapley value is equal to 0:

$$\varphi(\underline{x}_*, j) = 0.$$

- *Additivity*: if model $f()$ is a sum of two other models $g()$ and $h()$, then the Shapley value calculated for model $f()$ is a sum of Shapley values for models $g()$ and $h()$.

- *Local accuracy* (see Section 6.3.2): the sum of Shapley values is equal to the model's prediction, that is,

$$f(\underline{x}_*) - E_{\underline{X}}\{f(\underline{X})\} = \sum_{j=1}^{p} \varphi(\underline{x}_*, j),$$

where $\underline{X}$ is the vector of explanatory variables (corresponding to $\underline{x}_*$) that are treated as random values.

8.3   Example: Titanic data

Let us consider the random forest model `titanic_rf` (see Section 4.2.2) and passenger Johnny D (see Section 4.2.5) as the instance of interest in the Titanic data.

Box plots in Figure 8.3 present the distribution of the contributions $\Delta^{j|\pi(J,j)}(\underline{x}_*)$ for each explanatory variable of the model for 25 random orderings of the explanatory variables. Red and green bars represent, respectively, the negative and positive Shapley values across the orderings. It is clear that the young age of Johnny D results in a positive contribution for all orderings; the resulting Shapley value is equal to 0.2525. On the other hand, the effect of gender is in all cases negative, with the Shapley value equal to -0.0908.

The picture for variables *fare* and *class* is more complex, as their contributions can even change the sign, depending on the ordering. Note that Figure 8.3 presents Shapley values separately for each of the two variables. However, it is worth recalling that the iBD plot in Figure 7.1 indicated an important contribution of an interaction between the two variables. Hence, their contributions should not be separated. Thus, the Shapley values for *fare* and *class*, presented in Figure 8.3, should be interpreted with caution.

FIGURE 8.3 Explanatory-variable attributions for the prediction for Johnny D for the random forest model `titanic_rf` and the Titanic data based on 25 random orderings. Left-hand-side plot: box plots summarize the distribution of the attributions for each explanatory variable across the orderings. Red and green bars present Shapley values. Right-hand-side plot: Shapley values (mean attributions) without box plots.

In most applications, the detailed information about the distribution of variable contributions across the considered orderings of explanatory variables may not be of interest. Thus, one could simplify the plot by presenting only the Shapley values, as illustrated in the right-hand-side panel of Figure 8.3. Table 8.1 presents the Shapley values underlying this plot.

TABLE 8.1: Shapley values for the prediction for Johnny D for the random forest model `titanic_rf` and the Titanic data based on 25 random orderings.

Variable	Shapley value
age = 8	0.2525
class = 1st	0.0246
embarked = Southampton	-0.0032
fare = 72	0.0140
gender = male	-0.0943
parch = 0	-0.0097
sibsp = 0	0.0027

8.4   Pros and cons

Shapley values provide a uniform approach to decompose a model's predictions into contributions that can be attributed additively to different explanatory variables. Lundberg and Lee (2017) showed that the method unifies different approaches to additive variable attributions, like DeepLIFT (Shrikumar et al., 2017), Layer-Wise Relevance Propagation (Binder et al., 2016), or Local Interpretable Model-agnostic Explanations (Ribeiro et al., 2016). The method has got a strong formal foundation derived from the cooperative games theory. It also enjoys an efficient implementation in Python, with ports or re-implementations in R.

An important drawback of Shapley values is that they provide additive contributions (attributions) of explanatory variables. If the model is not additive, then the Shapley values may be misleading. This issue can be seen as arising from the fact that, in cooperative games, the goal is to distribute the payoff among payers. However, in the predictive modelling context, we want to understand how do the players affect the payoff? Thus, we are not limited to independent payoff-splits for players.

It is worth noting that, for an additive model, the approaches presented in Chapters 6–7 and in the current one lead to the same attributions. The reason is that, for additive models, different orderings lead to the same contributions. Since Shapley values can be seen as the mean across all orderings, it is essentially an average of identical values, i.e., it also assumes the same value.

An important practical limitation of the general model-agnostic method is that, for large models, the calculation of Shapley values is time-consuming. However, sub-sampling can be used to address the issue. For tree-based models, effective implementations are available.

8.5 Code snippets for R

In this section, we use the `DALEX` package, which is a wrapper for the `iBreakDown` R package. The package covers all methods presented in this chapter. It is available on `CRAN` and `GitHub`. Note that there are also other R packages that offer similar functionalities, like `iml` (Molnar et al., 2018), `fastshap` (Greenwell, 2020) or `shapper` (Maksymiuk et al., 2019), which is a wrapper for the Python library `SHAP` (Lundberg, 2019).

For illustration purposes, we use the `titanic_rf` random forest model for the Titanic data developed in Section 4.2.2. Recall that the model is developed to predict the probability of survival for passengers of Titanic. Instance-level explanations are calculated for Henry, a 47-year-old passenger that travelled in the first class (see Section 4.2.5).

We first retrieve the `titanic_rf` model-object and the data frame for Henry via the `archivist` hooks, as listed in Section 4.2.7. We also retrieve the version of the `titanic` data with imputed missing values.

```
titanic_imputed <- archivist::aread("pbiecek/models/27e5c")
titanic_rf <- archivist:: aread("pbiecek/models/4e0fc")
henry <- archivist::aread("pbiecek/models/a6538")
```

Then we construct the explainer for the model by using function `explain()` from the `DALEX` package (see Section 4.2.6). We also load the `randomForest` package, as the model was fitted by using function `randomForest()` from this package (see Section 4.2.2) and it is important to have the corresponding `predict()` function available. The model's prediction for Henry is obtained with the help of the function.

```
library("randomForest")
library("DALEX")
explain_rf <- DALEX::explain(model = titanic_rf,
                        data = titanic_imputed[, -9],
                           y = titanic_imputed$survived == "yes",
                       label = "Random Forest")
predict(explain_rf, henry)
```

```
## [1] 0.246
```

To compute Shapley values for Henry, we apply function `predict_parts()` (as in Section 6.6) to the explainer-object `explain_rf` and the data frame for the instance of interest, i.e., Henry. By specifying the `type="shap"` argument we indicate that we want to compute Shapley values. Additionally, the B=25 argument indicates that we want to select 25 random orderings of explanatory variables for which the Shapley values are to be computed. Note that B=25 is also the default value of the argument.

```
shap_henry <- predict_parts(explainer = explain_rf,
                      new_observation = henry,
                                 type = "shap",
                                    B = 25)
```

The resulting object `shap_henry` is a data frame with variable-specific attributions computed for every ordering. Printing out the object provides various summary statistics of the attributions including, of course, the mean.

```
shap_henry
```

```
##                                              min          q1
## Random Forest: age = 47               -0.14872225 -0.081197100
## Random Forest: class = 1st             0.12112732  0.123195061
## Random Forest: embarked = Cherbourg    0.01245129  0.022680335
## Random Forest: fare = 25              -0.03180517 -0.011710693
## Random Forest: gender = male          -0.15670412 -0.145184866
## Random Forest: parch = 0              -0.02795650 -0.007438151
## Random Forest: sibsp = 0              -0.03593203 -0.012978704
```

```
##                                    median          mean
## Random Forest: age = 47          -0.040909832 -0.060137381
## Random Forest: class = 1st        0.159974789  0.159090494
## Random Forest: embarked = Cherbourg 0.045746262  0.051056420
## Random Forest: fare = 25         -0.008647485  0.002175261
## Random Forest: gender = male     -0.126003135 -0.126984069
## Random Forest: parch = 0         -0.003043951 -0.005439239
## Random Forest: sibsp = 0         -0.005466244 -0.009070956
##                                       q3           max
## Random Forest: age = 47          -0.0230765745 -0.004967830
## Random Forest: class = 1st        0.1851354780  0.232307204
## Random Forest: embarked = Cherbourg 0.0558871772  0.117857725
## Random Forest: fare = 25          0.0162267784  0.070487540
## Random Forest: gender = male     -0.1115160852 -0.101295877
## Random Forest: parch = 0         -0.0008337109  0.003412778
## Random Forest: sibsp = 0          0.0031207522  0.007650204
```

By applying the generic function `plot()` to the `shap_henry` object, we obtain a graphical illustration of the results.

```
plot(shap_henry)
```

The resulting plot is shown in Figure 8.4. It includes the Shapley values and box plots summarizing the distributions of the variable-specific contributions for the selected random orderings.

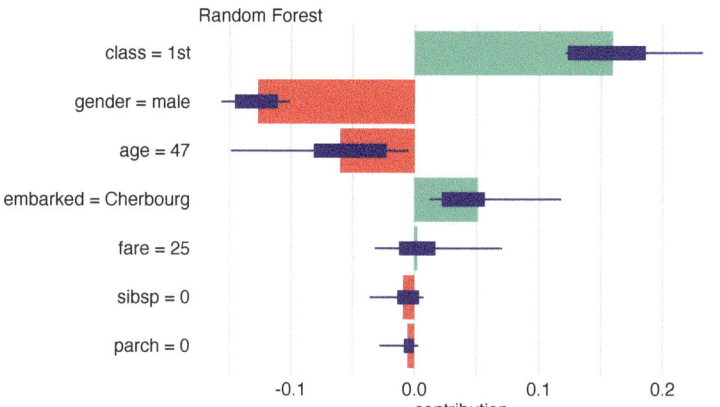

FIGURE 8.4 A plot of Shapley values with box plots for the `titanic_rf` model and passenger Henry for the Titanic data, obtained by applying the generic `plot()` function in R.

To obtain a plot with only Shapley values, i.e., without the box plots, we apply the `show_boxplots=FALSE` argument in the `plot()` function call.

```
plot(shap_henry, show_boxplots = FALSE)
```

The resulting plot, shown in Figure 8.5, can be compared to the plot in the right-hand-side panel of Figure 8.3 for Johnny D. The most remarkable difference is related to the contribution of *age*. The young age of Johnny D markedly increases the chances of survival, contrary to the negative contribution of the age of 47 for Henry.

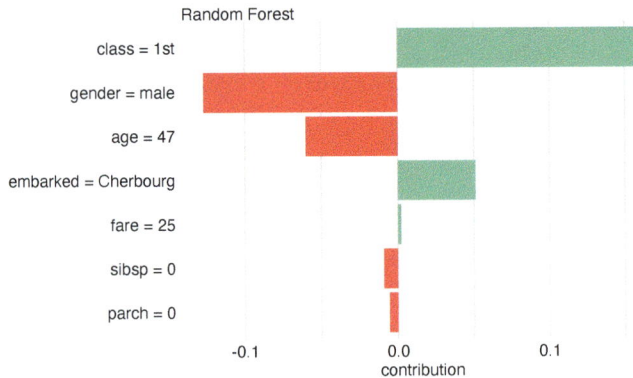

FIGURE 8.5 A plot of Shapley values without box plots for the `titanic_rf` model and passenger Henry for the Titanic data, obtained by applying the generic `plot()` function in R.

8.6 Code snippets for Python

In this section, we use the `dalex` library for Python. The package covers all methods presented in this chapter. It is available on `pip` and `GitHub`. Note that the most popular implementation in Python is available in the `shap` library (Lundberg and Lee, 2017). In this section, however, we show implementations from the `dalex` library because they are consistent with other methods presented in this book.

For illustration purposes, we use the `titanic_rf` random forest model for the Titanic data developed in Section 4.3.2. Instance-level explanations are calculated for Henry, a 47-year-old passenger that travelled in the 1st class (see Section 4.3.5).

In the first step, we create an explainer-object that provides a uniform interface for the predictive model. We use the `Explainer()` constructor for this purpose (see Section 4.3.6).

```
import pandas as pd
henry = pd.DataFrame({'gender'   : ['male'],
                      'age'      : [47],
                      'class'    : ['1st'],
                      'embarked': ['Cherbourg'],
                      'fare'     : [25],
                      'sibsp'    : [0],
                      'parch'    : [0]},
                     index = ['Henry'])

import dalex as dx
titanic_rf_exp = dx.Explainer(titanic_rf, X, y,
                  label = "Titanic RF Pipeline")
```

To calculate Shapley values we use the `predict_parts()` method with the `type='shap'` argument (see Section 6.7). The first argument indicates the data observation for which the values are to be calculated. Results are stored in the `results` field.

```
bd_henry = titanic_rf_exp.predict_parts(henry, type = 'shap')
bd_henry.result
```

	variable	contribution	variable_name	variable_value	sign	label	B
0	embarked = Cherbourg	0.022906	embarked	Cherbourg	1.0	Titanic RF Pipeline	1
1	fare = 25.0	0.006197	fare	25	1.0	Titanic RF Pipeline	1
2	class = 1st	0.076625	class	1st	1.0	Titanic RF Pipeline	1
3	gender = male	-0.105446	gender	male	-1.0	Titanic RF Pipeline	1
4	parch = 0.0	-0.005913	parch	0	-1.0	Titanic RF Pipeline	1
...	...	...	...	...	...	...	...
0	embarked = Cherbourg	0.005873	embarked	Cherbourg	1.0	Titanic RF Pipeline	0
1	fare = 25.0	-0.012377	fare	25	-1.0	Titanic RF Pipeline	0
3	gender = male	-0.002577	gender	male	-1.0	Titanic RF Pipeline	0
4	parch = 0.0	-0.007007	parch	0	-1.0	Titanic RF Pipeline	0
6	sibsp = 0.0	-0.013843	sibsp	0	-1.0	Titanic RF Pipeline	0

To visualize the obtained values, we simply call the `plot()` method.

```
bd_henry.plot()
```

The resulting plot is shown in Figure 8.6.

By default, Shapley values are calculated and plotted for all variables in the data. To limit the number of variables included in the graph, we can use the argument `max_vars` in the `plot()` method (see Section 6.7).

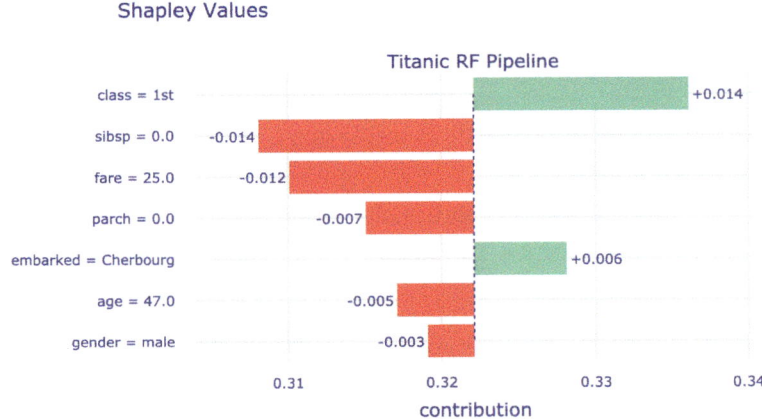

FIGURE 8.6 A plot of Shapley values for the `titanic_rf` model and passenger Henry for the Titanic data, obtained by applying the `plot()` method in Python.

9

Local Interpretable Model-agnostic Explanations (LIME)

9.1 Introduction

Break-down (BD) plots and Shapley values, introduced in Chapters 6 and 8, respectively, are most suitable for models with a small or moderate number of explanatory variables.

None of those approaches is well-suited for models with a very large number of explanatory variables, because they usually determine non-zero attributions for all variables in the model. However, in domains like, for instance, genomics or image recognition, models with hundreds of thousands, or even millions, of explanatory (input) variables are not uncommon. In such cases, sparse explanations with a small number of variables offer a useful alternative. The most popular example of such sparse explainers is the Local Interpretable Model-agnostic Explanations (LIME) method and its modifications.

The LIME method was originally proposed by Ribeiro et al. (2016). The key idea behind it is to locally approximate a black-box model by a simpler glass-box model, which is easier to interpret. In this chapter, we describe this approach.

9.2 Intuition

The intuition behind the LIME method is explained in Figure 9.1. We want to understand the factors that influence a complex black-box model around a single instance of interest (black cross). The coloured areas presented in Figure 9.1 correspond to decision regions for a binary classifier, i.e., they pertain to a prediction of a value of a binary dependent variable. The axes represent the values of two continuous explanatory variables. The coloured areas indicate combinations of values of the two variables for which the model classifies the observation to one of the two classes. To understand the local behavior of the complex model around the point of interest, we generate an artificial dataset, to which we fit a glass-box model. The dots in Figure 9.1 represent the generated artificial data; the size of the dots corresponds to proximity to the instance of interest. We can fit a simpler glass-box model to the artificial data so that it will locally approximate the predictions of the black-box model. In Figure 9.1, a simple linear model (indicated by the dashed line) is used to construct the local approximation. The simpler model serves as a "local explainer" for the more complex model.

We may select different classes of glass-box models. The most typical choices are regularized linear models like LASSO regression (Tibshirani, 1994) or decision trees (Hothorn et al., 2006). Both lead to sparse models that are easier to understand. The important point is to limit the complexity of the models, so that they are easier to explain.

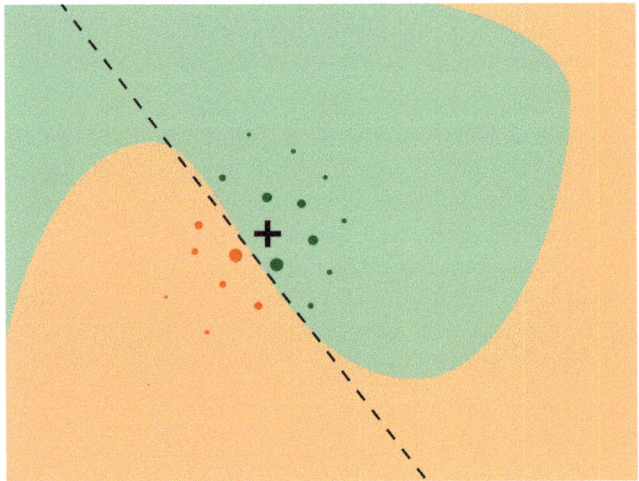

FIGURE 9.1 The idea behind the LIME approximation with a local glass-box model. The coloured areas correspond to decision regions for a complex binary classification model. The black cross indicates the instance (observation) of interest. Dots correspond to artificial data around the instance of interest. The dashed line represents a simple linear model fitted to the artificial data. The simple model "explains" local behavior of the black-box model around the instance of interest.

9.3   Method

We want to find a model that locally approximates a black-box model $f()$ around the instance of interest $\underline{x}_*$. Consider class G of simple, interpretable models like, for instance, linear models or decision trees. To find the required approximation, we minimize a "loss function":

$$\hat{g} = \arg \min_{g \in \mathcal{G}} L\{f, g, \nu(\underline{x}_*)\} + \Omega(g),$$

where model $g()$ belongs to class $\mathcal{G}$, $\nu(\underline{x}_*)$ defines a neighborhood of $\underline{x}_*$ in which approximation is sought, $L()$ is a function measuring the discrepancy between models $f()$ and $g()$ in the neighborhood $\nu(\underline{x}_*)$, and $\Omega(g)$ is a penalty for the complexity of model $g()$. The penalty is used to favour simpler models from class $\mathcal{G}$. In applications, this criterion is very often simplified by limiting class G to models with the same complexity, i.e., with the same number of coefficients. In such a situation, $\Omega(g)$ is the same for each model $g()$, so it can be omitted in optimization.

Note that models $f()$ and $g()$ may operate on different data spaces. The black-box model (function) $f(\underline{x}) : \mathcal{X} \to \mathcal{R}$ is defined on a large, p-dimensional space $\mathcal{X}$ corresponding to the p explanatory variables used in the model. The glass-box model (function) $g(\underline{x}) : \tilde{\mathcal{X}} \to \mathcal{R}$ is defined on a q-dimensional space $\tilde{\mathcal{X}}$ with $q << p$, often called the "space for interpretable representation". We will present some examples of $\tilde{\mathcal{X}}$ in the next section. For now we will just assume that some function $h()$ transforms $\mathcal{X}$ into $\tilde{\mathcal{X}}$.

If we limit class $\mathcal{G}$ to linear models with a limited number, say K, of non-zero coefficients, then the following algorithm may be used to find an interpretable glass-box model $g()$ that includes K most important, interpretable, explanatory variables:

```
Input: x* - observation to be explained
Input: N  - sample size for the glass-box model
Input: K  - complexity, the number of variables for the glass-box model
Input: similarity - a distance function in the original data space
1. Let x' = h(x*) be a version of x* in the lower-dimensional space
2. for i in 1...N {
3.    z'[i] <- sample_around(x')
4.    y'[i] <- f(z'[i])        # prediction for new observation z'[i]
5.    w'[i] <- similarity(x', z'[i])
6. }
7. return K-LASSO(y', x', w')
```

In Step 7, K-LASSO(y', x', w') stands for a weighted LASSO linear-regression that selects K variables based on the new data y' and x' with weights w'.

Practical implementation of this idea involves three important steps, which
are discussed in the subsequent subsections.

9.3.1   Interpretable data representation

As it has been mentioned, the black-box model $f()$ and the glass-box model $g()$
operate on different data spaces. For example, let us consider a VGG16 neural
network (Simonyan and Zisserman, 2015) trained on the ImageNet data (Deng
et al., 2009). The model uses an image of the size of 244×244 pixels as input
and predicts to which of 1000 potential categories does the image belong to.
The original space $\mathcal{X}$ is of dimension $3 \times 244 \times 244$ (three single-color channels
(*red, green, blue*) for a single pixel $\times$ 244 $\times$ 244 pixels), i.e., the input space is
178,608-dimensional. Explaining predictions in such a high-dimensional space is
difficult. Instead, from the perspective of a single instance of interest, the space
can be transformed into superpixels, which are treated as binary features that
can be turned on or off. Figure 9.2 (right-hand-side panel) presents an example
of 100 superpixels created for an ambiguous picture. Thus, in this case the
black-box model $f()$ operates on space $\mathcal{X} = \mathcal{R}^{178608}$, while the glass-box model
$g()$ applies to space $\tilde{\mathcal{X}} = \{0, 1\}^{100}$.

It is worth noting that superpixels, based on image segmentation, are frequent
choices for image data. For text data, groups of words are frequently used
as interpretable variables. For tabular data, continuous variables are often
discretized to obtain interpretable categorical data. In the case of categorical
variables, combination of categories is often used. We will present examples in
the next section.

FIGURE 9.2 The left-hand-side panel shows an ambiguous picture, half-horse
and half-duck (source Twitter https://bit.ly/3pJQ8eZ). The right-hand-side
panel shows 100 superpixels identified for this figure.

9.3.2 Sampling around the instance of interest

To develop a local-approximation glass-box model, we need new data points in the low-dimensional interpretable data space around the instance of interest. One could consider sampling the data points from the original dataset. However, there may not be enough points to sample from, because the data in high-dimensional datasets are usually very sparse and data points are "far" from each other. Thus, we need new, artificial data points. For this reason, the data for the development of the glass-box model is often created by using perturbations of the instance of interest.

For binary variables in the low-dimensional space, the common choice is to switch (from 0 to 1 or from 1 to 0) the value of a randomly-selected number of variables describing the instance of interest.

For continuous variables, various proposals have been formulated in different papers. For example, Molnar et al. (2018) and Molnar (2019) suggest adding Gaussian noise to continuous variables. Pedersen and Benesty (2019) propose to discretize continuous variables by using quantiles and then perturb the discretized versions of the variables. Staniak et al. (2019) discretize continuous variables based on segmentation of local ceteris-paribus profiles (for more information about the profiles, see Chapter 10).

In the example of the duck-horse image in Figure 9.2, the perturbations of the image could be created by randomly excluding some of the superpixels. An illustration of this process is shown in Figure 9.3.

FIGURE 9.3 The original image (left-hand-side panel) is transformed into a lower-dimensional data space by defining 100 super pixels (panel in the middle). The artificial data are created by using subsets of superpixels (right-hand-side panel).

9.3.3 Fitting the glass-box model

Once the artificial data around the instance of interest have been created, we may attempt to fit an interpretable glass-box model $g()$ from class $\mathcal{G}$.

The most common choices for class $\mathcal{G}$ are generalized linear models. To get sparse models, i.e., models with a limited number of variables, LASSO (least absolute shrinkage and selection operator) (Tibshirani, 1994) or similar regularization-modelling techniques are used. For instance, in the algorithm presented in Section 9.3, the K-LASSO method with K non-zero coefficients has been mentioned. An alternative choice are classification-and-regression trees models (Breiman et al., 1984).

For the example of the duck-horse image in Figure 9.2, the VGG16 network provides 1000 probabilities that the image belongs to one of the 1000 classes used for training the network. It appears that the two most likely classes for the image are *'standard poodle'* (probability of 0.18) and *'goose'* (probability of 0.15). Figure 9.4 presents LIME explanations for these two predictions. The explanations were obtained with the K-LASSO method, which selected $K = 15$ superpixels that were the most influential from a model-prediction point of view. For each of the selected two classes, the K superpixels with non-zero coefficients are highlighted. It is interesting to observe that the superpixel which contains the beak is influential for the *'goose'* prediction, while superpixels linked with the white colour are influential for the *'standard poodle'* prediction. At least for the former, the influential feature of the plot does correspond to the intended content of the image. Thus, the results of the explanation increase confidence in the model's predictions.

Label: standard poodle
Probability: 0.18
Explanation Fit: 0.37

Label: goose
Probability: 0.15
Explanation Fit: 0.55

FIGURE 9.4 LIME for two predictions ('standard poodle' and 'goose') obtained by the VGG16 network with ImageNet weights for the half-duck, half-horse image.

9.4 Example: Titanic data

Most examples of the LIME method are related to the text or image data. In this section, we present an example of a binary classification for tabular data to facilitate comparisons between methods introduced in different chapters.

Let us consider the random forest model `titanic_rf` (see Section 4.2.2) and passenger Johnny D (see Section 4.2.5) as the instance of interest for the Titanic data.

First, we have got to define an interpretable data space. One option would be to gather similar variables into larger constructs corresponding to some concepts. For example *class* and *fare* variables can be combined into "wealth", *age* and *gender* into "demography", and so on. In this example, however, we have got a relatively small number of variables, so we will use a simpler data representation in the form of a binary vector. Toward this aim, each variable is dichotomized into two levels. For example, *age* is transformed into a binary variable with categories "≤ 15.36" and ">15.36", *class* is transformed into a binary variable with categories "1st/2nd/deck crew" and "other", and so on. Once the lower-dimension data space is defined, the LIME algorithm is applied to this space. In particular, we first have got to appropriately transform data for Johnny D. Subsequently, we generate a new artificial dataset that will be used for K-LASSO approximations of the random forest model. In particular, the K-LASSO method with $K = 3$ is used to identify the three most influential (binary) variables that will provide an explanation for the prediction for Johnny D. The three variables are: *age*, *gender*, and *class*. This result agrees with the conclusions drawn in the previous chapters. Figure 9.5 shows the coefficients estimated for the K-LASSO model.

9.5 Pros and cons

As mentioned by Ribeiro et al. (2016), the LIME method

- is *model-agnostic*, as it does not imply any assumptions about the black-box model structure;
- offers an *interpretable representation*, because the original data space is transformed (for instance, by replacing individual pixels by superpixels for image data) into a more interpretable, lower-dimension space;
- provides *local fidelity*, i.e., the explanations are locally well-fitted to the black-box model.

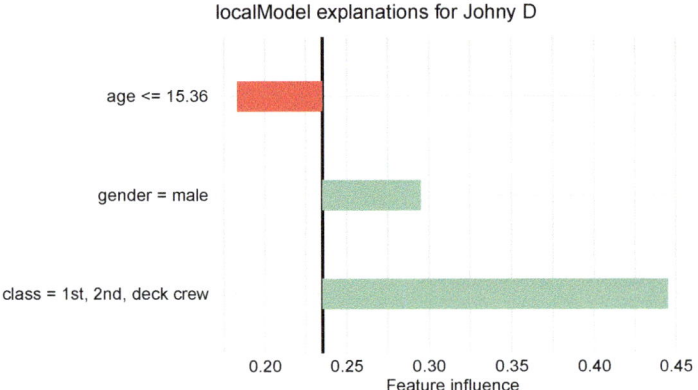

FIGURE 9.5 LIME method for the prediction for Johnny D for the random forest model `titanic_rf` and the Titanic data. Presented values are the coefficients of the K-LASSO model fitted locally to the predictions from the original model.

The method has been widely adopted in the text and image analysis, partly due to the interpretable data representation. In that case, the explanations are delivered in the form of fragments of an image/text, and users can easily find the justification of such explanations. The underlying intuition for the method is easy to understand: a simpler model is used to approximate a more complex one. By using a simpler model, with a smaller number of interpretable explanatory variables, predictions are easier to explain. The LIME method can be applied to complex, high-dimensional models.

There are several important limitations, however. For instance, as mentioned in Section 9.3.2, there have been various proposals for finding interpretable representations for continuous and categorical explanatory variables in case of tabular data. The issue has not been solved yet. This leads to different implementations of LIME, which use different variable-transformation methods and, consequently, that can lead to different results.

Another important point is that, because the glass-box model is selected to approximate the black-box model, and not the data themselves, the method does not control the quality of the local fit of the glass-box model to the data. Thus, the latter model may be misleading.

Finally, in high-dimensional data, data points are sparse. Defining a "local neighborhood" of the instance of interest may not be straightforward. Importance of the selection of the neighborhood is discussed, for example, by Alvarez-Melis and Jaakkola (2018). Sometimes even slight changes in the neighborhood strongly affect the obtained explanations.

To summarize, the most useful applications of LIME are limited to high-dimensional data for which one can define a low-dimensional interpretable data representation, as in image analysis, text analysis, or genomics.

9.6 Code snippets for R

LIME and its variants are implemented in various R and Python packages. For example, `lime` (Pedersen and Benesty, 2019) started as a port of the LIME Python library (Lundberg, 2019), while `localModel` (Staniak et al., 2019), and `iml` (Molnar et al., 2018) are separate packages that implement a version of this method entirely in R.

Different implementations of LIME offer different algorithms for extraction of interpretable features, different methods for sampling, and different methods of weighting. For instance, regarding transformation of continuous variables into interpretable features, `lime` performs global discretization using quartiles, `localModel` performs local discretization using ceteris-paribus profiles (for more information about the profiles, see Chapter 10), while `iml` works directly on continuous variables. Due to these differences, the packages yield different results (explanations).

Also, `lime`, `localModel`, and `iml` use different functions to implement the LIME method. Thus, we will use the `predict_surrogate()` method from the `localModel` package. The function offers a uniform interface to the functions from the three packages.

For illustration purposes, we use the `titanic_rf` random forest model for the Titanic data developed in Section 4.2.2. Recall that it is developed to predict the probability of survival from the sinking of the Titanic. Instance-level explanations are calculated for Johnny D, an 8-year-old passenger that travelled in the first class. We retrieve the `titanic_rf` model-object and the data frame for Johnny D via the `archivist` hooks, as listed in Section 4.2.7. We also retrieve the version of the `titanic` data with imputed missing values.

```
titanic_imputed <- archivist::aread("pbiecek/models/27e5c")
titanic_rf <- archivist:: aread("pbiecek/models/4e0fc")
johnny_d <- archivist:: aread("pbiecek/models/e3596")
```

```
  class gender age sibsp parch fare     embarked
1   1st   male   8     0     0   72 Southampton
```

Then we construct the explainer for the model by using the function `explain()` from the `DALEX` package (see Section 4.2.6). We also load the `randomForest` package, as the model was fitted by using function `randomForest()` from this package (see Section 4.2.2) and it is important to have the corresponding `predict()` function available.

```
library("randomForest")
library("DALEX")
titanic_rf_exp <- DALEX::explain(model = titanic_rf,
                      data = titanic_imputed[, -9],
                         y = titanic_imputed$survived == "yes",
                     label = "Random Forest")
```

9.6.1   The lime package

The key functions in the `lime` package are `lime()`, which creates an explanation, and `explain()`, which evaluates explanations. As mentioned earlier, we will apply the `predict_surrogate()` function from the `DALEXtra` package to access the functions via an interface that is consistent with the approach used in the previous chapters.

The `predict_surrogate()` function requires two arguments: `explainer`, which specifies the name of the explainer-object created with the help of function `explain()` from the `DALEX` package, and `new_observation`, which specifies the name of the data frame for the instance for which prediction is of interest. An additional, important argument is `type` that indicates the package with the desired implementation of the LIME method: either `"localModel"` (default), `"lime"`, or `"iml"`. In case of the `lime`-package implementation, we can specify two additional arguments: `n_features` to indicate the maximum number (K) of explanatory variables to be selected by the K-LASSO method, and `n_permutations` to specify the number of artifical data points to be sampled for the local-model approximation.

In the code below, we apply the `predict_surrogate()` function to the explainer-object for the random forest model `titanic_rf` and data for Johnny D. Additionally, we specify that the K-LASSO method should select no more than `n_features=3` explanatory variables based on a fit to `n_permutations=1000` sampled data points. Note that we use the `set.seed()` function to ensure repeatability of the sampling.

```
set.seed(1)
library("DALEXtra")
library("lime")

lime_johnny <- predict_surrogate(explainer = titanic_rf_exp,
                  new_observation = johnny_d,
                  n_features = 3,
                  n_permutations = 1000,
                  type = "lime")
```

The contents of the resulting object can be printed out in the form of a data frame with 11 variables.

```
(as.data.frame(lime_johnny))
```

```
##    model_type case  model_r2 model_intercept model_prediction feature
## 1  regression    1 0.6826437       0.5541115        0.4784804  gender
## 2  regression    1 0.6826437       0.5541115        0.4784804     age
## 3  regression    1 0.6826437       0.5541115        0.4784804   class
##    feature_value feature_weight  feature_desc                        data
## 1              2     -0.4038175 gender = male 1, 2, 8, 0, 0, 72, 4
## 2              8      0.1636630     age <= 22 1, 2, 8, 0, 0, 72, 4
## 3              1      0.1645234   class = 1st 1, 2, 8, 0, 0, 72, 4
##    prediction
## 1       0.422
## 2       0.422
## 3       0.422
```

The output includes column `case` that provides indices of observations for which the explanations are calculated. In our case there is only one index equal to 1, because we asked for an explanation for only one observation, Johnny D. The `feature` column indicates which explanatory variables were given non-zero coefficients in the K-LASSO method. The `feature_value` column provides information about the values of the original explanatory variables for the observations for which the explanations are calculated. On the other hand, the `feature_desc` column indicates how the original explanatory variable was transformed. Note that the applied implementation of the LIME method dichotomizes continuous variables by using quartiles. Hence, for instance, *age* for Johnny D was transformed into a binary variable `age <= 22`.

Column `feature_weight` provides the estimated coefficients for the variables selected by the K-LASSO method for the explanation. The `model_intercept` column provides of the value of the intercept. Thus, the linear combination of the transformed explanatory variables used in the glass-box model approximating the random forest model around the instance of interest, Johnny D, is given by the following equation (see Section 2.5):

$$\hat{p}_{lime} = 0.55411 - 0.40381 \cdot 1_{male} + 0.16366 \cdot 1_{age<=22} + 0.16452 \cdot 1_{class=1st} = 0.47848,$$

where 1_A denotes the indicator variable for condition A. Note that the computed value corresponds to the number given in the column `model_prediction` in the printed output.

By applying the `plot()` function to the object containing the explanation, we obtain a graphical presentation of the results.

```
plot(lime_johnny)
```

The resulting plot is shown in Figure 9.6. The length of the bar indicates the magnitude (absolute value), while the color indicates the sign (red for negative, blue for positive) of the estimated coefficient.

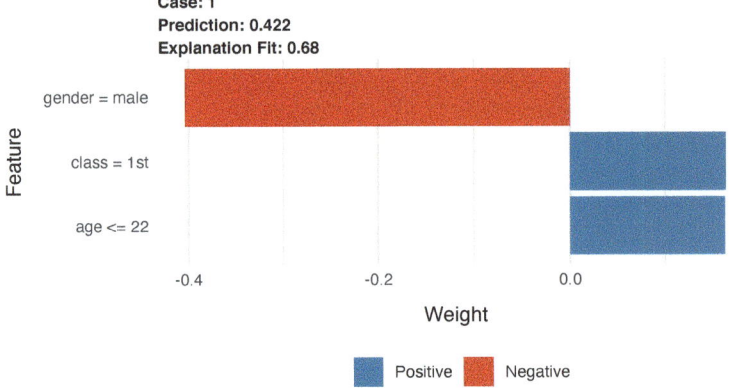

FIGURE 9.6 Illustration of the LIME-method results for the prediction for Johnny D for the random forest model `titanic_rf` and the Titanic data, generated by the `lime` package.

9.6.2 The `localModel` package

The key function of the `localModel` package is the `individual_surrogate_model()` function that fits the local glass-box model. The function is applied to the explainer-object obtained with the help of the `DALEX::explain()` function (see Section 4.2.6). As mentioned earlier, we will apply the `predict_surrogate()` function from the `DALEXtra` package to access the functions via an interface that is consistent with the approach used in the previous chapters. To choose the `localModel`-implementation of LIME, we set argument `type="localMode"` (see Section 9.6.1). In that case, the method accepts, apart from the required arguments `explainer` and `new_observation`, two additional arguments: `size`, which specifies the number of artificial data points to be sampled for the local-model approximation, and `seed`, which sets the seed for the random-number generation allowing for a repeatable execution.

In the code below, we apply the `predict_surrogate()` function to the explainer-object for the random forest model `titanic_rf` and data for Johnny D. Additionally, we specify that 1000 data points are to be sampled and we set the random-number-generation seed.

```
library("localModel")
locMod_johnny <- predict_surrogate(explainer = titanic_rf_exp,
                    new_observation = johnny_d,
                    size = 1000,
                    seed = 1,
                    type = "localModel")
```

The resulting object is a data frame with seven variables (columns). For brevity, we only print out the first three variables.

```
locMod_johnny[,1:3]
```

```
##       estimated                    variable original_variable
## 1   0.23530947                (Model mean)
## 2   0.30331646                 (Intercept)
## 3   0.06004988            gender = male              gender
## 4  -0.05222505              age <= 15.36                 age
## 5   0.20988506     class = 1st, 2nd, deck crew            class
## 6   0.00000000 embarked = Belfast, Southampton         embarked
```

The printed output includes column `estimated` that contains the estimated coefficients of the LASSO regression model, which is used to approximate the predictions from the random forest model. Column `variable` provides the information about the corresponding variables, which are transformations of `original_variable`. Note that the version of LIME, implemented in the `localModel` package, dichotomizes continuous variables by using ceteris-paribus profiles (for more information about the profiles, see Chapter 10). The profile for variable *age* for Johnny D can be obtained by using function `plot_interpretable_feature()`, as shown below.

```
plot_interpretable_feature(locMod_johnny, "age")
```

The resulting plot is presented in Figure 9.7. The profile indicates that the largest drop in the predicted probability of survival is observed when the value of *age* increases beyond about 15 years. Hence, in the output of the `predict_surrogate()` function, we see a binary variable `age <= 15.36`, as Johnny D was 8 years old.

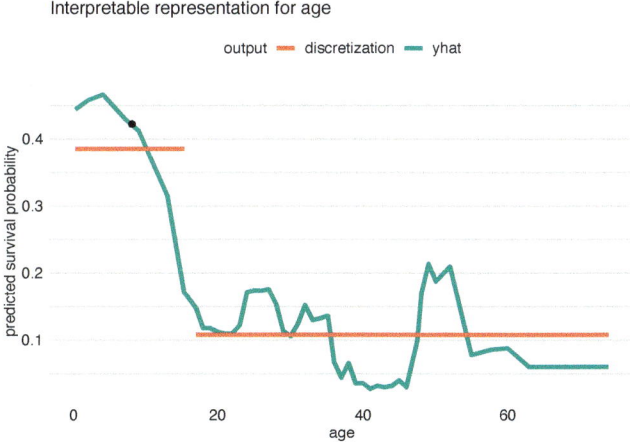

FIGURE 9.7 Discretization of the *age* variable for Johnny D based on the ceteris-paribus profile. The optimal change-point is around 15 years of age.

By applying the generic `plot()` function to the object containing the LIME-method results, we obtain a graphical representation of the results.

```
plot(locMod_johnny)
```

The resulting plot is shown in Figure 9.8. The lengths of the bars indicate the magnitude (absolute value) of the estimated coefficients of the LASSO logistic regression model. The bars are placed relative to the value of the mean prediction, 0.235.

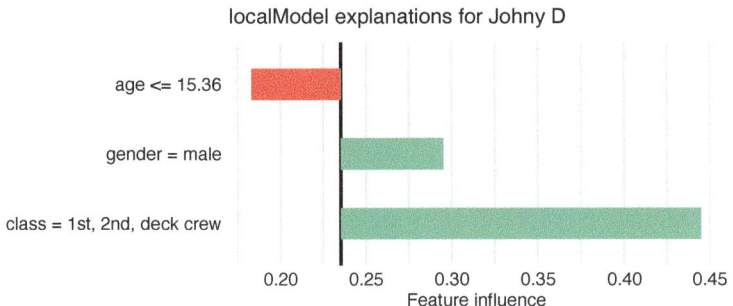

FIGURE 9.8 Illustration of the LIME-method results for the prediction for Johnny D for the random forest model `titanic_rf` and the Titanic data, generated by the `localModel` package.

9.6.3   The `iml` package

The key functions of the `iml` package are `Predictor$new()`, which creates an explainer, and `LocalModel$new()`, which develops the local glass-box model. The main arguments of the `Predictor$new()` function are `model`, which specifies the model-object, and `data`, the data frame used for fitting the model. As mentioned earlier, we will apply the `predict_surrogate()` function from the `DALEXtra` package to access the functions via an interface that is consistent with the approach used in the previous chapters. To choose the `iml`-implementation of LIME, we set argument `type="iml"` (see Section 9.6.1). In that case, the method accepts, apart from the required arguments `explainer`and `new_observation`, an additional argument `k` that specifies the number of explanatory variables included in the local-approximation model.

```
library("DALEXtra")
library("iml")
iml_johnny <- predict_surrogate(explainer = titanic_rf_exp,
                new_observation = johnny_d,
                k = 3,
                type = "iml")
```

The resulting object includes data frame `results` with seven variables that provides results of the LASSO logistic regression model which is used to approximate the predictions of the random forest model. For brevity, we print out selected variables.

```
iml_johnny$results[,c(1:5,7)]
```

```
##                 beta x.recoded       effect x.original     feature .class
## 1 -0.1992616770         1 -0.19926168        1st    class=1st     no
## 2  1.6005493672         1  1.60054937       male gender=male     no
## 3 -0.0002111346        72 -0.01520169         72         fare     no
## 4  0.1992616770         1  0.19926168        1st    class=1st    yes
## 5 -1.6005493672         1 -1.60054937       male gender=male    yes
## 6  0.0002111346        72  0.01520169         72         fare    yes
```

The printed output includes column `beta` that provides the estimated coefficients of the local-approximation model. Note that two sets of three coefficients (six in total) are given, corresponding to the prediction of the probability of death (column `.class` assuming value `no`, corresponding to the value `"no"` of the `survived` dependent-variable) and survival (`.class` asuming value `yes`). Column `x.recoded` contains the information about the value of the corresponding transformed (interpretable) variable. The value of the original explanatory variable is given in column `x.original`, with column `feature` providing the information about the corresponding variable. Note that the implemented version of LIME does not transform continuous variables. Categorical variables are dichotomized, with the resulting binary variable assuming the value of 1 for the category observed for the instance of interest and 0 for other categories.

The `effect` column provides the product of the estimated coefficient (from column `beta`) and the value of the interpretable covariate (from column `x.recoded`) of the model approximating the random forest model.

By applying the generic `plot()` function to the object containing the LIME-method results, we obtain a graphical representation of the results.

```
plot(iml_johnny)
```

The resulting plot is shown in Figure 9.9. It shows values of the two sets of three coefficients for both types of predictions (probability of death and survival).

```
plot(iml_johnny)
```

It is worth noting that *age*, *gender*, and *class* are correlated. For instance, crew members are only adults and mainly men. This is probably the reason why the three packages implementing the LIME method generate slightly different explanations for the model prediction for Johnny D.

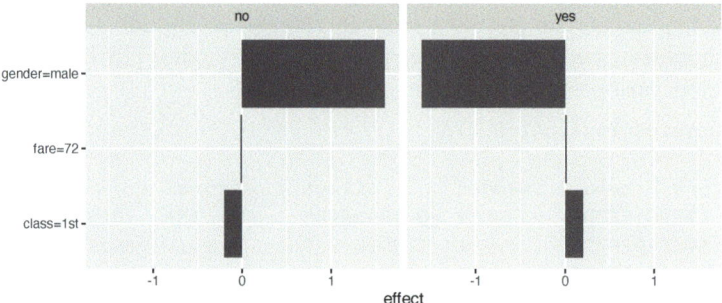

FIGURE 9.9 Illustration of the LIME-method results for the prediction for Johnny D for the random forest model `titanic_rf` and the Titanic data, generated by the `iml` package.

9.7   Code snippets for Python

In this section, we use the `lime` library for Python, which is probably the most popular implementation of the LIME method (Ribeiro et al., 2016). The `lime` library requires categorical variables to be encoded in a numerical format. This requires some additional work with the data. Therefore, below we will show you how to use this method in Python step by step.

For illustration purposes, we use the random forest model for the Titanic data. Instance-level explanations are calculated for Henry, a 47-year-old passenger that travelled in the 1st class.

In the first step, we read the Titanic data and encode categorical variables. In this case, we use the simplest encoding for *gender*, *class*, and *embarked*, i.e., the label-encoding.

```python
import dalex as dx

titanic = dx.datasets.load_titanic()
X = titanic.drop(columns='survived')
y = titanic.survived

from sklearn import preprocessing
le = preprocessing.LabelEncoder()

X['gender']   = le.fit_transform(X['gender'])
X['class']    = le.fit_transform(X['class'])
X['embarked'] = le.fit_transform(X['embarked'])
```

In the next step we train a random forest model.

```
from sklearn.ensemble import RandomForestClassifier as rfc
titanic_fr = rfc()
titanic_fr.fit(X, y)
```

It is time to define the observation for which model prediction will be explained. We write Henry's data into `pandas.Series` object.

```
import pandas as pd
henry = pd.Series([1, 47.0, 0, 1, 25.0, 0, 0],
                  index =['gender', 'age', 'class', 'embarked',
                          'fare', 'sibsp', 'parch'])
```

The `lime` library explains models that operate on images, text, or tabular data. In the latter case, we have to use the `LimeTabularExplainer` module.

```
from lime.lime_tabular import LimeTabularExplainer
explainer = LimeTabularExplainer(X,
                  feature_names=X.columns,
                  class_names=['died', 'survived'],
                  discretize_continuous=False,
                  verbose=True)
```

The result is an explainer that can be used to interpret a model around specific observations. In the following example, we explain the behaviour of the model for Henry. The `explain_instance()` method finds a local approximation with an interpretable linear model. The result can be presented graphically with the `show_in_notebook()` method.

```
lime = explainer.explain_instance(henry, titanic_fr.predict_proba)
lime.show_in_notebook(show_table=True)
```

The resulting plot is shown in Figure 9.10.

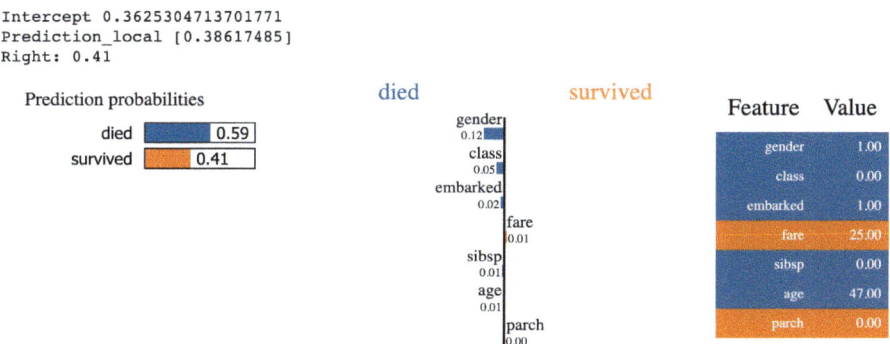

FIGURE 9.10 A plot of LIME model values for the random forest model and passenger Henry for the Titanic data.

10

Ceteris-paribus Profiles

10.1 Introduction

Chapters 6–9 focused on the methods that quantified the importance of explanatory variables in the context of a single-instance prediction. Their application yields a decomposition of the prediction into components that can be attributed to particular variables. In this chapter, we focus on a method that evaluates the effect of a selected explanatory variable in terms of changes of a model's prediction induced by changes in the variable's values. The method is based on the *ceteris-paribus* principle. *"Ceteris paribus"* is a Latin phrase meaning "other things held constant" or "all else unchanged". The method examines the influence of an explanatory variable by assuming that the values of all other variables do not change. The main goal is to understand how changes in the values of the variable affect the model's predictions.

Explanation tools (explainers) presented in this chapter are linked to the second law introduced in Section 1.2, i.e., the law of "Prediction's speculation". This is why the tools are also known as "What-if" model analysis or Individual Conditional Expectations (Goldstein et al., 2015). It appears that it is easier to understand how a black-box model works if we can explore the model by investigating the influence of explanatory variables separately, changing one at a time.

10.2 Intuition

Ceteris-paribus (CP) profiles show how a model's prediction would change if the value of a single exploratory variable changed. In essence, a CP profile shows the dependence of the conditional expectation of the dependent variable (response) on the values of the particular explanatory variable. For example, panel A of Figure 10.1 presents response (prediction) surface for two explanatory variables, *age* and *class*, for the logistic regression model `titanic_lmr` (see Section 4.2.1) for the Titanic dataset (see Section 4.1). We are interested in the change of the model's prediction for passenger Henry (see Section 4.2.5) induced by each of the variables. Toward this end, we may want to explore the curvature of the response surface around a single point with *age* equal to 47 and *class* equal to "1st", indicated in the plot. CP profiles are one-dimensional plots that examine the curvature across each dimension, i.e., for each variable. Panel B of Figure 10.1 presents the CP profiles for *age* and *class*. Note that, in the CP profile for *age*, the point of interest is indicated by the dot. The plots for both variables suggest that the predicted probability of survival varies considerably for different ages and classes.

10.3   Method

In this section, we introduce more formally one-dimensional CP profiles. Recall (see Section 2.3) that we use $\underline{x}_i$ to refer to the vector of values of explanatory variables corresponding to the i-th observation in a dataset. A vector with arbitrary values (not linked to any particular observation in the dataset) is denoted by $\underline{x}_*$. Let x_*^j denote the j-th element of $\underline{x}_*$, i.e., the value of the j-th explanatory variable. We use $\underline{x}_*^{-j}$ to refer to a vector resulting from removing the j-th element from $\underline{x}_*$. By $\underline{x}_*^{j|=z}$, we denote a vector resulting from changing the value of the j-th element of $\underline{x}_*$ to (a scalar) z.

We define a one-dimensional CP profile $h()$ for model $f()$, the j-th explanatory variable, and point of interest $\underline{x}_*$ as follows:

$$h_{\underline{x}_*}^{f,j}(z) \equiv f\left(\underline{x}_*^{j|=z}\right). \tag{10.1}$$

CP profile is a function that describes the dependence of the (approximated) conditional expected value (prediction) of Y on the value z of the j-th explanatory variable. Note that, in practice, z assumes values from the entire observed range for the variable, while values of all other explanatory variables are kept fixed at the values specified by $\underline{x}_*$.

Note that, in the situation when only a single model is considered, we will skip the model index and we will denote the CP profile for the j-th explanatory variable and the point of interest $\underline{x}_*$ by $h_{\underline{x}_*}^j(z)$.

10.4   Example: Titanic data

For continuous explanatory variables, a natural way to represent the CP function (10.1) is to use a plot similar to one of those presented in Figure 10.2.

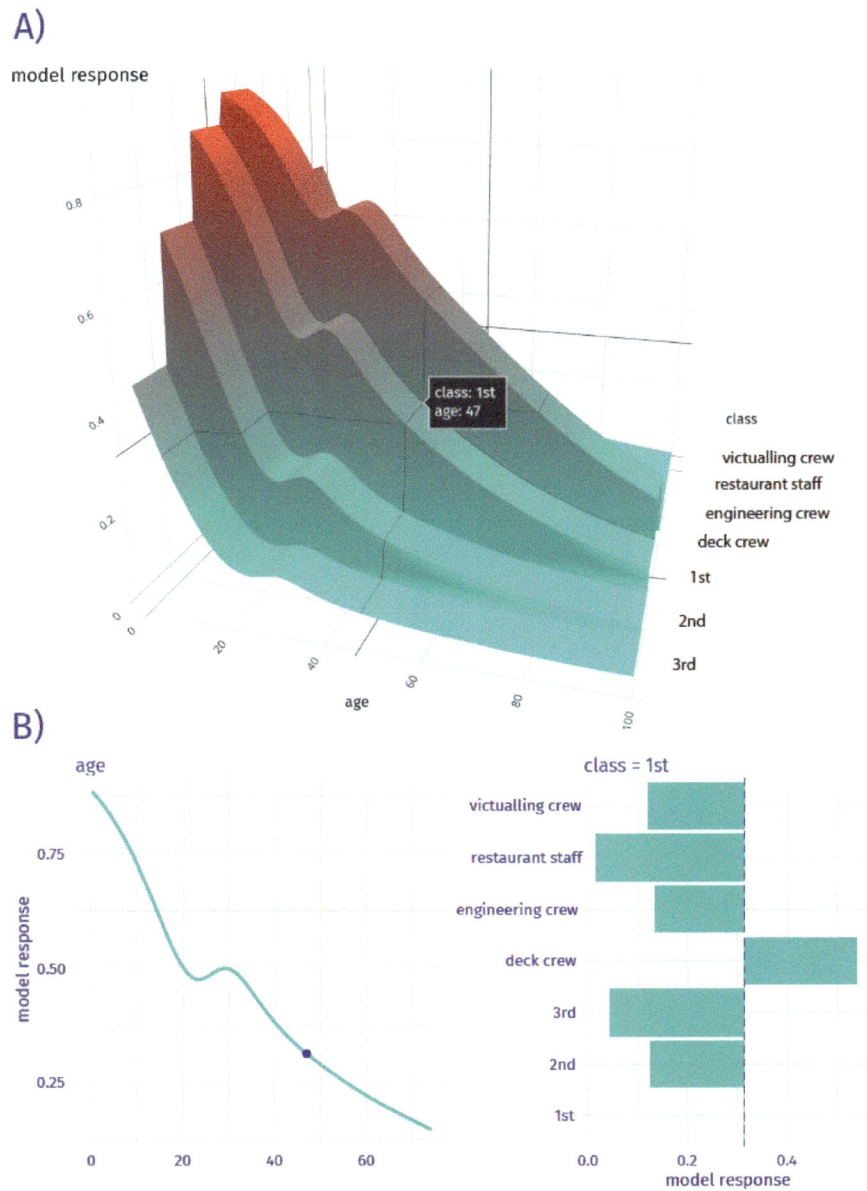

FIGURE 10.1 Panel A) shows the model response (prediction) surface for variables *age* and *class*. Ceteris-paribus (CP) profiles are conditional, one-dimensional plots that are marked with black curves. They help to understand the changes of the curvature of the surface induced by changes in only a single explanatory variable. Panel B) CP profiles for individual variables, *age* (continuous) and *class* (categorical).

In the figure, the dot on the curves marks the instance-prediction of interest, i.e., prediction $f(\underline{x}_*)$ for a single observation $\underline{x}_*$. The curve itself shows how the prediction would change if the value of a particular explanatory variable changed.

In particular, Figure 10.2 presents CP profiles for the *age* variable for the logistic regression model `titanic_lmr` and the random forest model `titanic_rf` for the Titanic dataset (see Sections 4.2.1 and 4.2.2, respectively). The instance of interest is passenger Henry, a 47-year-old man who travelled in the first class (see Section 4.2.5). It is worth observing that the profile for the logistic regression model is smooth, while the one for the random forest model is a step function with some variability. However, the general shape of the two CP profiles is similar. If Henry were a newborn, while keeping values of all other explanatory variables unchanged, his predicted survival probability would increase by about 40 percentage points for both models. And if Henry were 80 years old, the predictions would decrease by more than 10 percentage points.

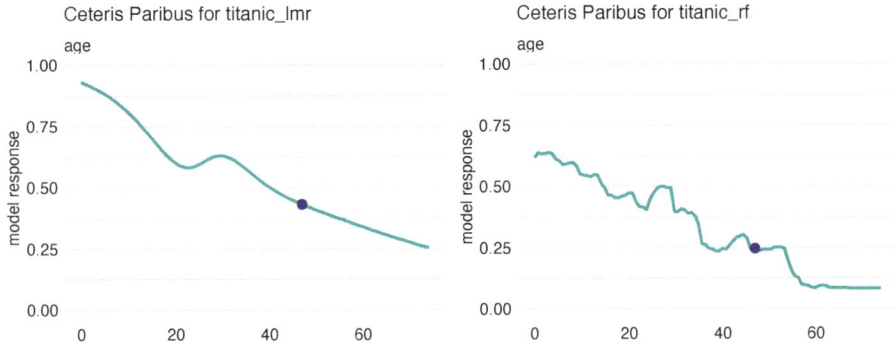

FIGURE 10.2 Ceteris-paribus profiles for variable *age* for the logistic regression (`titanic_lmr`) and random forest (`titanic_rf`) models that predict the probability of surviving of passenger Henry based on the Titanic data. Dots indicate the values of the variable and of the prediction for Henry.

For a categorical explanatory variable, a natural way to represent the CP function is to use a bar plot similar to one of those presented in Figure 10.3. In particular, the figure presents CP profiles for the *class* variable in the logistic regression and random forest models for the Titanic dataset (see Sections 4.2.1 and 4.2.2, respectively). For this instance (observation), passenger Henry, the predicted probability for the logistic regression model would decrease substantially if the value of *class* changed to "2nd" or "3rd". On the other hand, for the random forest model, the largest change would be marked if *class* changed to "desk crew".

Usually, black-box models contain a large number of explanatory variables. However, CP profiles are legible even for tiny subplots, if created with techniques like sparklines or small multiples (Tufte, 1986). By using the techniques,

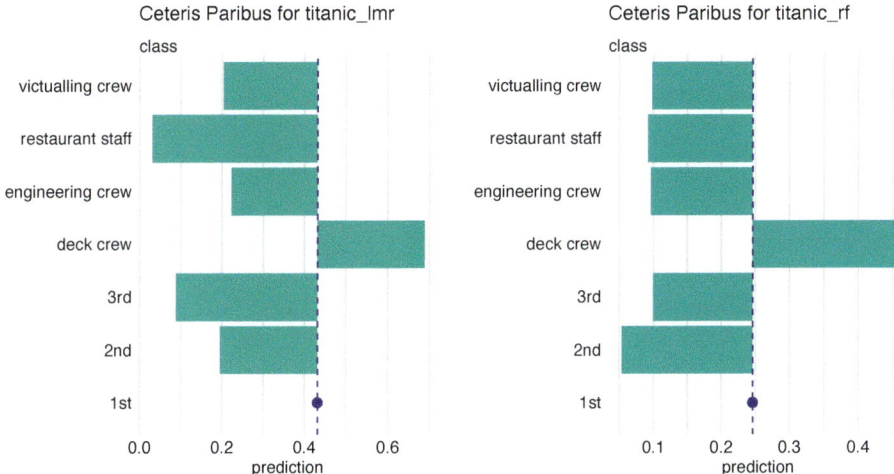

FIGURE 10.3 Ceteris-paribus profiles for variable *class* for the logistic regression (`titanic_lmr`) and random forest (`titanic_rf`) models that predict the probability of surviving of passenger Henry based on the Titanic data. Dots indicate the values of the variable and of the prediction for Henry.

we can display a large number of profiles, while at the same time keeping profiles for consecutive variables in separate panels, as shown in Figure 10.4 for the random forest model for the Titanic dataset. It helps if the panels are ordered so that the most important profiles are listed first. A method to assess the importance of CP profiles is discussed in the next chapter.

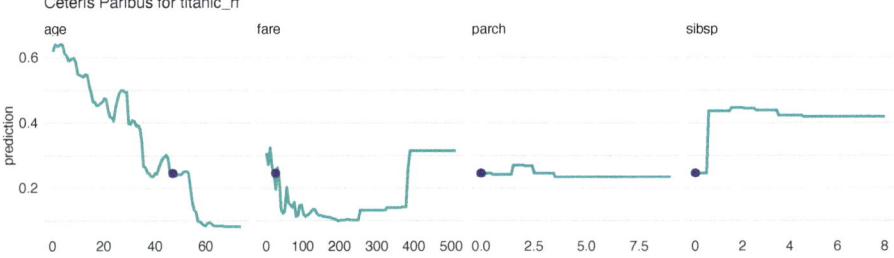

FIGURE 10.4 Ceteris-paribus profiles for all continuous explanatory variables for the random forest model **titanic_rf** for the Titanic dataset and passenger Henry. Dots indicate the values of the variables and of the prediction for Henry.

10.5   Pros and cons

One-dimensional CP profiles, as presented in this chapter, offer a uniform, easy to communicate, and extendable approach to model exploration. Their graphical representation is easy to understand and explain. It is possible to show profiles for many variables or models in a single plot. CP profiles are easy to compare, as we can overlay profiles for two or more models to better understand differences between the models. We can also compare two or more instances to better understand model-prediction's stability. CP profiles are also a useful tool for sensitivity analysis.

However, there are several issues related to the use of the CP profiles. One of the most important ones is related to the presence of correlated explanatory variables. For such variables, the application of the *ceteris-paribus* principle may lead to unrealistic settings and misleading results, as it is not possible to keep one variable fixed while varying the other one. For example, variables like surface and number of rooms, which can be used in prediction of an apartment's price, are usually correlated. Thus, it is unrealistic to consider very small apartments with a large number of rooms. In fact, in a training dataset, there may be no such combinations. Yet, as implied by (10.1), to compute a CP profile for the number-of-rooms variable for a particular instance of a small-surface apartment, we should consider the model's predictions $f\left(\underline{x}_*^{j|=z}\right)$ for all values of z (i.e., numbers of rooms) observed in the training dataset, including large ones. This means that, especially for flexible models like, for example, regression trees, predictions for a large number of rooms z may have to be obtained by extrapolating the results obtained for large-surface apartments. Needless to say, such extrapolation may be problematic. We will come back to this issue in Chapters 17 and 18.

A somewhat similar issue is related to the presence of interactions in a model, as they imply the dependence of the effect of one variable on other one(s). Pairwise interactions require the use of two-dimensional CP profiles that are more complex than one-dimensional ones. Clearly, interactions of higher orders pose even a greater challenge.

A practical issue is that, in case of a model with hundreds or thousands of variables, the number of plots to inspect may be daunting.

Finally, while bar plots allow visualization of CP profiles for factors (categorical explanatory variables), their use becomes less trivial in case of factors with many nominal (unordered) categories (like, for example, a ZIP-code).

10.6 Code snippets for R

In this section, we present CP profiles as implemented in the **DALEX** package for R. Note that presented functions are, in fact, wrappers to package **ingredients** (Biecek et al., 2019) with a simplified interface. There are also other R packages that offer similar functionalities, like **condvis** (O'Connell et al., 2017), **pdp** (Greenwell, 2017), **ICEbox** (Goldstein et al., 2015), **ALEPlot** (Apley, 2018), or **iml** (Molnar et al., 2018).

For illustration, we use two classification models developed in Chapter 4.1, namely the logistic regression model **titanic_lmr** (Section 4.2.1) and the random forest model **titanic_rf** (Section 4.2.2). They are developed to predict the probability of survival after sinking of Titanic. Instance-level explanations are calculated for Henry, a 47-year-old male passenger that travelled in the first class (see Section 4.2.5).

We first retrieve the **titanic_lmr** and **titanic_rf** model-objects and the data frame for Henry via the **archivist** hooks, as listed in Section 4.2.7. We also retrieve the version of the **titanic** data with imputed missing values.

```
titanic_imputed <- archivist::aread("pbiecek/models/27e5c")
titanic_lmr <- archivist::aread("pbiecek/models/58b24")
titanic_rf <- archivist::aread("pbiecek/models/4e0fc")
(henry <- archivist::aread("pbiecek/models/a6538"))
```

```
  class gender age sibsp parch fare  embarked
1 1st   male   47  0     0     25 Cherbourg
```

Then we construct the explainers for the model by using function `explain()` from the **DALEX** package (see Section 4.2.6). We also load the **rms** and **randomForest** packages, as the models were fitted by using functions from those packages and it is important to have the corresponding `predict()` functions available.

```
library("DALEX")
library("rms")
explain_lmr <- explain(model = titanic_lmr,
                       data  = titanic_imputed[, -9],
                       y     = titanic_imputed$survived == "yes",
                       type = "classification",
                       label = "Logistic Regression")

library("randomForest")
explain_rf <- DALEX::explain(model = titanic_rf,
                       data  = titanic_imputed[, -9],
                       y     = titanic_imputed$survived == "yes",
                       label = "Random Forest")
```

10.6.1   Basic use of the `predict_profile()` function

The easiest way to create and plot CP profiles is to use the `predict_profile()` function and then apply the generic `plot()` function to the resulting object. By default, profiles for all explanatory variables are calculated, while profiles for all numeric (continuous) variables are plotted. One can limit the number of variables for which calculations and/or plots are necessary by using the `variables` argument.

To compute the CP profiles, the `predict_profile()` function requires arguments `explainer`, which specifies the name of the explainer-object, and `new_observation`, which specifies the name of the data frame for the instance for which prediction is of interest. As a result, the function returns an object of class "ceteris_paribus_explainer". It is a data frame with the model's predictions. Below we illustrate the use of the function for the random forest model.

```
cp_titanic_rf <- predict_profile(explainer = explain_rf,
                         new_observation = henry)
cp_titanic_rf
```

```
## Top profiles    :
##                  class gender age sibsp parch fare   embarked _yhat_
## 1                  1st   male  47     0     0   25  Cherbourg  0.246
## 1.1                2nd   male  47     0     0   25  Cherbourg  0.054
## 1.2                3rd   male  47     0     0   25  Cherbourg  0.100
## 1.3          deck crew   male  47     0     0   25  Cherbourg  0.454
## 1.4 engineering crew    male  47     0     0   25  Cherbourg  0.096
## 1.5 restaurant staff    male  47     0     0   25  Cherbourg  0.092
##      _vname_ _ids_        _label_
## 1      class     1 Random Forest
## 1.1    class     1 Random Forest
## 1.2    class     1 Random Forest
## 1.3    class     1 Random Forest
## 1.4    class     1 Random Forest
## 1.5    class     1 Random Forest
##
##
## Top observations:
##   class gender age sibsp parch fare   embarked _yhat_       _label_
## 1   1st   male  47     0     0   25  Cherbourg  0.246 Random Forest
##   _ids_
## 1     1
```

To obtain a graphical representation of CP profiles, the generic `plot()` function can be applied to the data frame returned by the `predict_profile()` function. It returns a `ggplot2` object that can be processed further if needed. In the examples below, we use the `ggplot2` functions like `ggtitle()` or `ylim()` to modify the plot's title or the range of the y-axis.

Below we show the code that can be used to create plots similar to those presented in the upper part of Figure 10.4. By default, the `plot()` function provides a graph with plots for all numerical variables. To limit the display to variables *age* and *fare*, the names of the variables are provided in the `variables` argument. The resulting plot is shown in Figure 10.5.

```
library("ggplot2")
plot(cp_titanic_rf, variables = c("age", "fare")) +
  ggtitle("Ceteris-paribus profile", "") + ylim(0, 0.8)
```

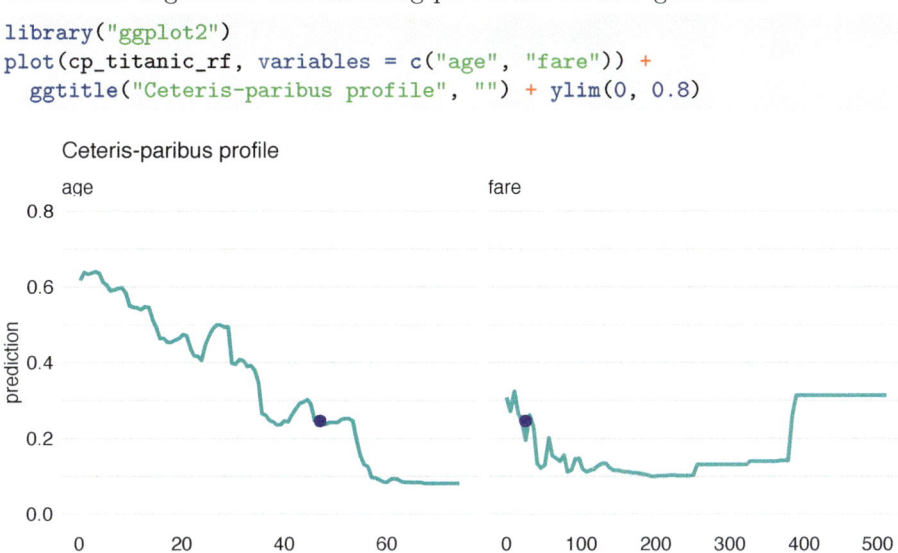

FIGURE 10.5 Ceteris-paribus profiles for variables *age* and *fare* and the `titanic_rf` random forest model for the Titanic data. Dots indicate the values of the variables and of the prediction for Henry.

To plot CP profiles for categorical variables, we have got to add the `variable_type = "categorical"` argument to the `plot()` function. In that case, we can use the `categorical_type` argument to specify whether we want to obtain a plot with `"lines"` (default) or `"bars"`. In the code below, we also use argument `variables` to indicate that we want to create plots only for variables *class* and *embarked*. The resulting plot is shown in Figure 10.6.

```
plot(cp_titanic_rf, variables = c("class", "embarked"),
     variable_type = "categorical", categorical_type = "bars") +
  ggtitle("Ceteris-paribus profile", "")
```

10.6.2   Advanced use of the `predict_profile()` function

The `predict_profile()` function is very flexible. To better understand how it can be used, we briefly review its arguments:

- `explainer, data, predict_function, label` - they provide information about the model. If the object provided in the `explainer` argument has been created with the `DALEX::explain()` function, then values of the other

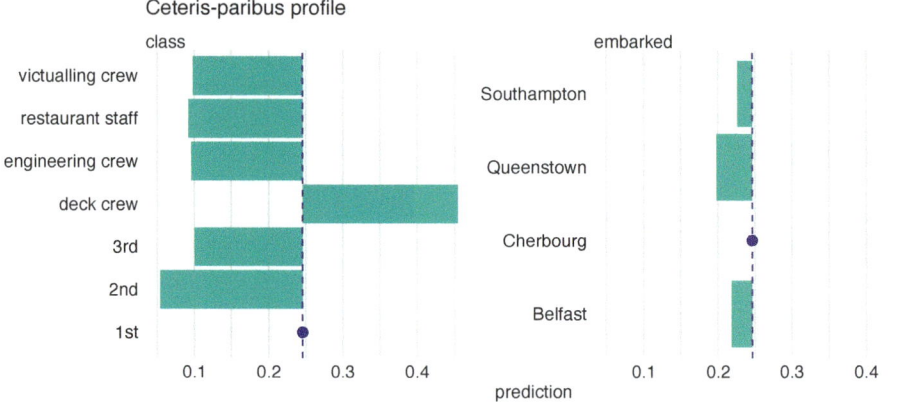

FIGURE 10.6 Ceteris-paribus profiles for variables *class* and *embarked* and the `titanic_rf` random forest model for the Titanic data. Dots indicate the values of the variables and of the prediction for Henry.

arguments are extracted from the object; this is how we use the function in this chapter. Otherwise, we have got to specify directly the model-object, the data frame used for fitting the model, the function that should be used to compute predictions, and the model label.

- `new_observation` - a data frame with data for instance(s), for which we want to calculate CP profiles, with the same variables as in the data used to fit the model. Note, however, that it is best not to include the dependent variable in the data frame, as they should not appear in plots.
- `y` - the observed values of the dependent variable corresponding to `new_observation`. The use of this argument is illustrated in Section 12.1.
- `variables` - names of explanatory variables, for which CP profiles are to be calculated. By default, `variables = NULL` and the profiles are constructed for all variables, which may be time consuming.
- `variable_splits` - a list of values for which CP profiles are to be calculated. By default, `variable_splits = NULL` and the list includes all values for categorical variables and uniformly-placed values for continuous variables; for the latter, one can specify the number of the values with the `grid_points` argument (by default, `grid_points = 101`).

The code below uses argument `variable_splits` to specify that CP profiles are to be calculated for *age* and *fare*, together with the list of values at which the profiles are to be evaluated.

```
variable_splits = list(age = seq(0, 70, 0.1),
                       fare = seq(0, 100, 0.1))
cp_titanic_rf <- predict_profile(explainer = explain_rf,
                                 new_observation = henry,
                                 variable_splits = variable_splits)
```

Subsequently, to replicate the plots presented in the upper part of Figure 10.4, a call to function `plot()` can be used as below. The resulting plot is shown in Figure 10.5.

```r
plot(cp_titanic_rf, variables = c("age", "fare")) +
  ggtitle("Ceteris-paribus profile", "")
```

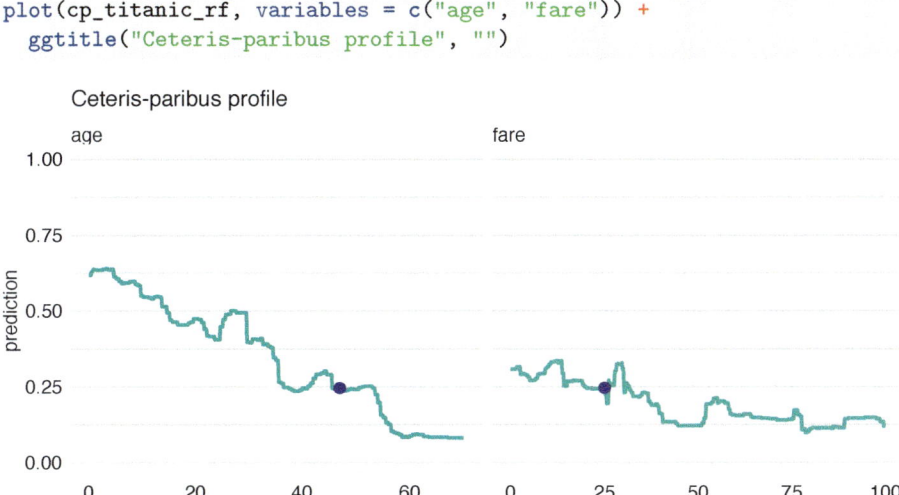

FIGURE 10.7 Ceteris-paribus profiles for variables *class* and *embarked* and the `titanic_rf` random forest model. Blue dots indicate the values of the variables and of the prediction for Henry.

In the example below, we present the code to create CP profiles for two passengers, Henry and Johnny D (see Section 4.2.5), for the random forest model `titanic_rf` (Section 4.2.2). Toward this end, we first retrieve the `johnny_d` data frame via the `archivist` hook, as listed in Section 4.2.7. We then apply the `predict_profile()` function with the explainer-object `explain_rf` specified in the `explainer` argument and the combined data frame for Henry and Johnny D used in the `new_observation` argument. We also use argument `variable_splits` to specify that CP profiles are to be calculated for *age* and *fare*, together with the list of values at which the profiles are to be evaluated.

```r
(johnny_d <- archivist::aread("pbiecek/models/e3596"))

##   class gender age sibsp parch fare   embarked
## 1   1st   male   8     0     0   72 Southampton

cp_titanic_rf2 <- predict_profile(explainer = explain_rf,
                         new_observation = rbind(henry, johnny_d),
                         variable_splits = variable_splits)
```

To create the plots of CP profile, we apply the `plot()` function. We use the `scale_color_manual()` function to add names of passengers to the plot, and to control colors and positions.

```
library(ingredients)
plot(cp_titanic_rf2, color = "_ids_", variables = c("age", "fare")) +
  scale_color_manual(name = "Passenger:", breaks = 1:2,
            values = c("#4378bf", "#8bdcbe"),
            labels = c("henry" , "johny_d"))
```

The resulting graph, which includes CP profiles for Henry and Johnny D, is presented in Figure 10.8. For Henry, the predicted probability of survival is smaller than for Johnny D, as seen from the location of the large dots on the profiles. The profiles for *age* indicate a somewhat larger effect of the variable for Henry, as the predicted probability, in general, decreases from about 0.6 to 0.1 with increasing values of the variable. For Johny D, the probability changes from about 0.45 to about 0.05, with a bit less monotonic pattern. For *fare*, the effect is smaller for both passengers, as the probability changes within a smaller range of about 0.2. For Henry, the changes are approximately limited to the interval [0.1,0.3], while for Johnny D they are limited to the interval [0.4,0.6].

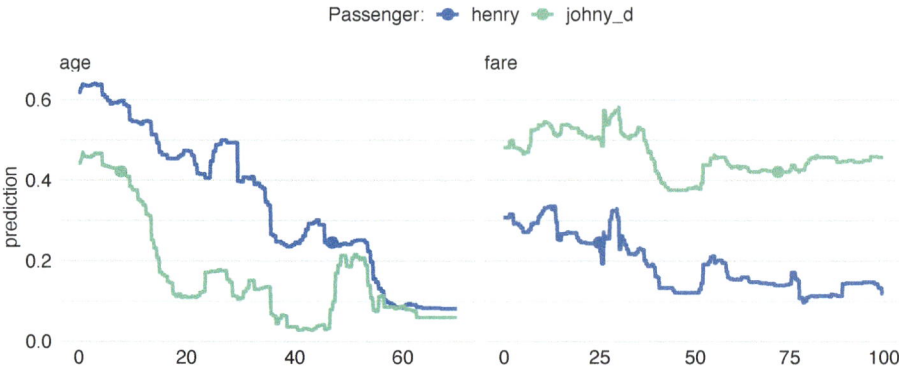

FIGURE 10.8 Ceteris-paribus profiles for the `titanic_rf` model. Profiles for different passengers are color-coded. Dots indicate the values of the variables and of the predictions for the passengers.

10.6.3   Comparison of models (champion-challenger)

One of the most interesting uses of the CP profiles is the comparison for two or more of models.

To illustrate this possibility, first, we have to construct profiles for the models. In our illustration, for the sake of clarity, we limit ourselves to the logistic regression (Section 4.2.1) and random forest (Section 4.2.2) models for the Titanic data. Moreover, we use Henry as the instance for which predictions are of interest. We apply the `predict_profile()` function to compute the CP profiles for the two models.

```
cp_titanic_rf <- predict_profile(explain_rf, henry)
cp_titanic_lmr <- predict_profile(explain_lmr, henry)
```

Subsequently, we construct the plot with the help of the `plot()` function. Note that, for the sake of brevity, we use the `variables` argument to limit the plot only to profiles for variables *age* and *class*. Every `plot()` function can take a collection of explainers as arguments. In such case, profiles for different models are combined in a single plot. In the code presented below, argument `color = "_label_"` is used to specify that models are to be color-coded. The `_label_` refers to the name of the column in the CP explainer that contains the model's name.

```
plot(cp_titanic_rf, cp_titanic_lmr, color = "_label_",
    variables = c("age", "fare")) +
    ggtitle("Ceteris-paribus profiles for Henry", "")
```

The resulting plot is shown in Figure 10.9. For Henry, the predicted probability of survival is higher for the logistic regression model than for the random forest model. CP profiles for *age* show a similar shape, however, and indicate decreasing probability with age. Note that this relation is not linear because we used spline transformation for the *age* variable, see Section 4.2.1. For *fare*, the profile for the logistic regression model suggests a slight increase of the probability, while for the random forest a decreasing trend can be inferred. The difference between the values of the CP profiles for *fare* increases with the increasing values of the variable. We can only speculate what is the reason for the difference. Perhaps the cause is the correlation between the ticket *fare* and *class*. The logistic regression model handles the dependency of variables differently from the random forest model.

10.7 Code snippets for Python

In this section, we use the `dalex` library for Python. The package covers all methods presented in this chapter. It is available on `pip` and `GitHub`.

For illustration purposes, we use the `titanic_rf` random forest model for the Titanic data developed in Section 4.3.2. Recall that the model is developed to predict the probability of survival for passengers of Titanic. Instance-level explanations are calculated for Henry, a 47-year-old passenger that travelled in the 1st class (see Section 4.3.5).

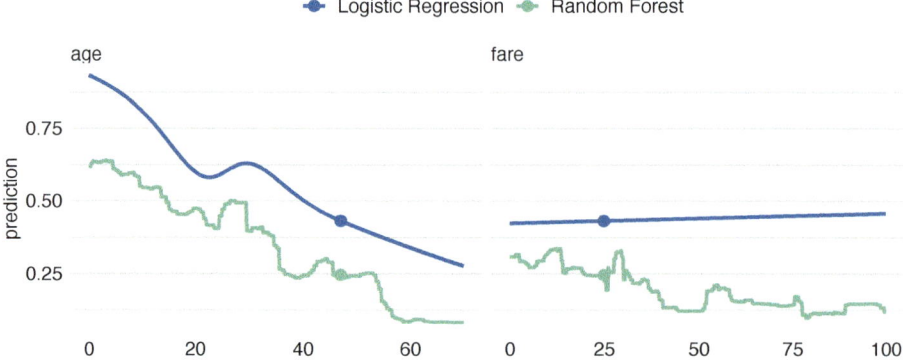

Ceteris-paribus profiles for Henry

FIGURE 10.9 Comparison of the ceteris-paribus profiles for Henry for the logistic regression and random forest models. Profiles for different models are color-coded. Dots indicate the values of the variables and of the prediction for Henry.

In the first step, we create an explainer-object that will provide a uniform interface for the predictive model. We use the `Explainer()` constructor for this purpose.

```
import pandas as pd
henry = pd.DataFrame({'gender'   : ['male'],
                      'age'      : [47],
                      'class'    : ['1st'],
                      'embarked' : ['Cherbourg'],
                      'fare'     : [25],
                      'sibsp'    : [0],
                      'parch'    : [0]},
                     index = ['Henry'])

import dalex as dx
titanic_rf_exp = dx.Explainer(titanic_rf, X, y,
                    label = "Titanic RF Pipeline")
```

To calculate the CP profile we use the `predict_profile()` method. The first argument is the data frame for the observation for which the attributions are to be calculated. Results are stored in the `results` field.

```
cp_henry = titanic_rf_exp.predict_profile(henry)
cp_henry.result
```

	gender	age	class	embarked	fare	sibsp	parch	_original_	_yhat_	_vname_	_ids_	_label_
Henry	male	47.000000	1st	Cherbourg	25.0	0.0	0.00	male	0.300822	gender	Henry	Titanic RF Pipeline
Henry	female	47.000000	1st	Cherbourg	25.0	0.0	0.00	male	0.815790	gender	Henry	Titanic RF Pipeline
Henry	male	0.166667	1st	Cherbourg	25.0	0.0	0.00	47	0.440580	age	Henry	Titanic RF Pipeline
Henry	male	0.905000	1st	Cherbourg	25.0	0.0	0.00	47	0.447339	age	Henry	Titanic RF Pipeline
Henry	male	1.643333	1st	Cherbourg	25.0	0.0	0.00	47	0.440729	age	Henry	Titanic RF Pipeline
...	...	...	...	...	...	...	...	...	...	...	...	...
Henry	male	47.000000	1st	Cherbourg	25.0	0.0	8.64	0	0.345057	parch	Henry	Titanic RF Pipeline
Henry	male	47.000000	1st	Cherbourg	25.0	0.0	8.73	0	0.345057	parch	Henry	Titanic RF Pipeline
Henry	male	47.000000	1st	Cherbourg	25.0	0.0	8.82	0	0.345057	parch	Henry	Titanic RF Pipeline
Henry	male	47.000000	1st	Cherbourg	25.0	0.0	8.91	0	0.345057	parch	Henry	Titanic RF Pipeline
Henry	male	47.000000	1st	Cherbourg	25.0	0.0	9.00	0	0.345057	parch	Henry	Titanic RF Pipeline

The resulting object can be visualised by using the `plot()` method. By default, CP profiles for all continuous variables are plotted. To select specific variables, a vector with the names of the variables can be provided in the `variables` argument. In the code below, we select variables *age* and *fare*. The resulting plot is shown in Figure 10.10.

```
cp_henry.plot(variables = ['age', 'fare'])
```

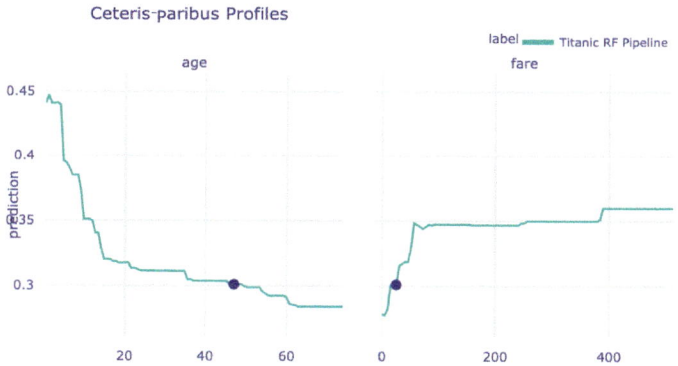

FIGURE 10.10 Ceteris-paribus profiles for continuous explanatory variables *age* and *fare* for the random forest model for the Titanic data and passenger Henry. Dots indicate the values of the variables and of the prediction for Henry.

To plot profiles for categorical variables, we use the `variable_type = 'categorical'` argument. In the code below, we limit the plot to variables *class* and *embarked*. The resulting plot is shown in Figure 10.11.

```
cp_henry.plot(variables = ['class', 'embarked'],
              variable_type = 'categorical')
```

CP profiles for several models can be placed on a single chart by adding them

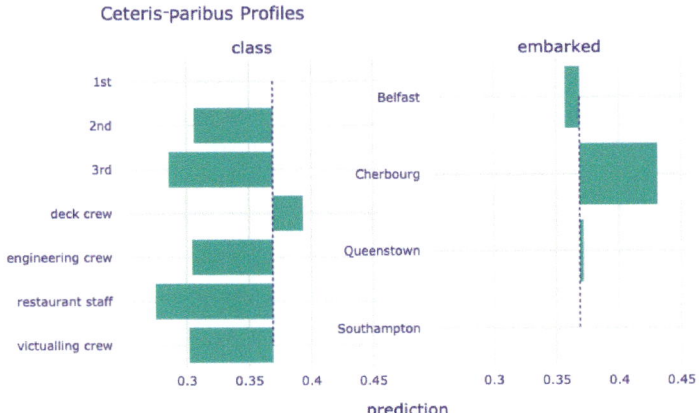

FIGURE 10.11 Ceteris-paribus profiles for categorical explanatory variables *class* and *embarked* for the random forest model for the Titanic data and passenger Henry.

as further arguments for the `plot()` function (see an example below). The resulting plot is shown in Figure 10.12.

```
cp_henry2 = titanic_lr_exp.predict_profile(henry)
cp_henry.plot(cp_henry2, variables = ['age', 'fare'])
```

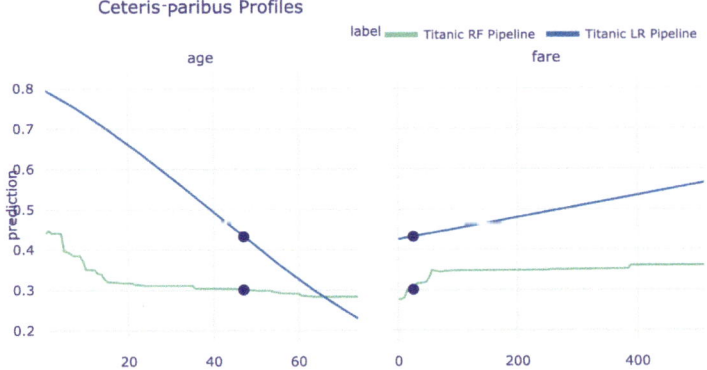

FIGURE 10.12 Ceteris-paribus profiles for logistic regression model and random forest model for the Titanic data and passenger Henry.

11

Ceteris-paribus Oscillations

11.1 Introduction

Visual examination of ceteris-paribus (CP) profiles, as illustrated in the previous chapter, is insightful. However, in case of a model with a large number of explanatory variables, we may end up with a large number of plots that may be overwhelming. In such a situation, it might be useful to select the most interesting or important profiles. In this chapter, we describe a measure that can be used for such a purpose and that is directly linked to CP profiles. It can be seen as an instance-level variable-importance measure alternative to the measures discussed in Chapters 6–9.

11.2 Intuition

To assign importance to CP profiles, we can use the concept of profile oscillations. It is worth noting that the larger influence of an explanatory variable on prediction for a particular instance, the larger the fluctuations of the corresponding CP profile. For a variable that exercises little or no influence on a model's prediction, the profile will be flat or will barely change. In other words, the values of the CP profile should be close to the value of the model's prediction for a particular instance. Consequently, the sum of differences between the profile and the value of the prediction, taken across all possible values of the explanatory variable, should be close to zero. The sum can be graphically depicted by the area between the profile and the horizontal line representing the value of the single-instance prediction. On the other hand, for an explanatory variable with a large influence on the prediction, the area should be large. Figure 11.1 illustrates the concept based on CP profiles presented in Figure 10.4. The larger the highlighted area in Figure 11.1, the more important is the variable for the particular prediction.

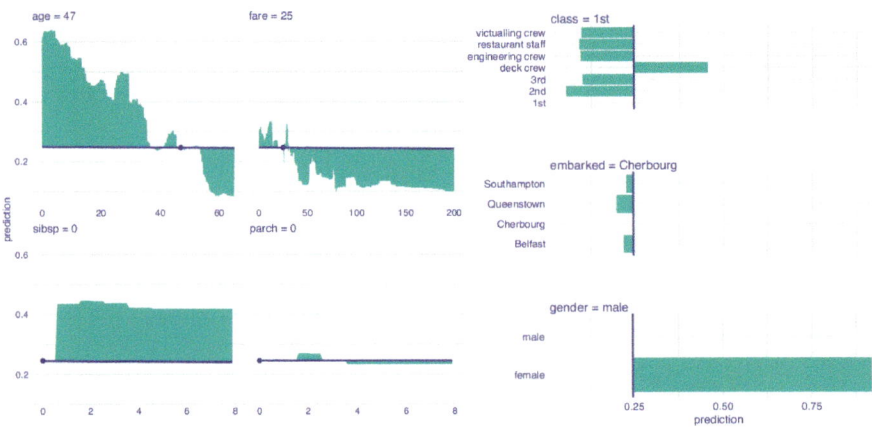

FIGURE 11.1 The value of the coloured area summarizes the oscillations of a ceteris-paribus (CP) profile and provides the mean of the absolute deviations between the CP profile and the single-instance prediction. The CP profiles are constructed for the `titanic_rf` random forest model for the Titanic data and passenger Henry.

11.3 Method

Let us formalize this concept now. Denote by $g^j(z)$ the probability density function of the distribution of the j-th explanatory variable. The summary measure of the variable's importance for model $f()$'s prediction at $\underline{x}_*$, $vip^j_{CP}(\underline{x}_*)$, is defined as follows:

$$vip^j_{CP}(\underline{x}_*) = \int_{\mathcal{R}} |h^j_{\underline{x}_*}(z) - f(\underline{x}_*)| g^j(z) dz = E_{X^j}\left\{|h^j_{\underline{x}_*}(X^j) - f(\underline{x}_*)|\right\}. \quad (11.1)$$

Thus, $vip^j_{CP}(\underline{x}_*)$ is the expected absolute deviation of the CP profile $h^j_{\underline{x}_*}()$, defined in (10.1), from the model's prediction at $\underline{x}_*$, computed over the distribution $g^j(z)$ of the j-th explanatory variable.

The true distribution of j-th explanatory variable is, in most cases, unknown. There are several possible approaches to construct an estimator of (11.1).

One is to calculate the area under the CP curve, i.e., to assume that $g^j(z)$ is a uniform distribution over the range of variable X^j. It follows that a straightforward estimator of $vip^j_{CP}(\underline{x}_*)$ is

$$\widehat{vip}_{CP}^{j,uni}(\underline{x}_*) = \frac{1}{k} \sum_{l=1}^{k} |h_{\underline{x}_*}^j(z_l) - f(\underline{x}_*)|, \qquad (11.2)$$

where z_l ($l = 1, \ldots, k$) are selected values of the j-th explanatory variable. For instance, one can consider all unique values of X^j in a dataset. Alternatively, for a continuous variable, one can use an equidistant grid of values.

Another approach is to use the empirical distribution of X^j. This leads to the estimator defined as follows:

$$\widehat{vip}_{CP}^{j,emp}(\underline{x}_*) = \frac{1}{n} \sum_{i=1}^{n} |h_{\underline{x}_*}^j(x_i^j) - f(\underline{x}_*)|, \qquad (11.3)$$

where index i runs through all observations in a dataset.

The use of $\widehat{vip}_{CP}^{j,emp}(\underline{x}_*)$ is preferred when there are enough data to accurately estimate the empirical distribution and when the distribution is not uniform. On the other hand, $\widehat{vip}_{CP}^{j,uni}(\underline{x}_*)$ is in most cases quicker to compute and, therefore, it is preferred if we look for fast approximations.

Note that the local evaluation of the variables' importance can be very different from the global evaluation. This is well illustrated by the following example. Consider the model

$$f(x^1, x^2) = x^1 \cdot x^2,$$

where variables X^1 and X^2 take values in $[0, 1]$. Furthermore, consider prediction for an observation described by vector $\underline{x}_* = (0, 1)$. In that case, the importance of X^1 is larger than X^2. This is because the CP profile $h_{\underline{x}_*}^1(z) = z$, while $h_{\underline{x}_*}^2(z) = 0$. Thus, there are oscillations for the first variable, but no oscillations for the second one. Hence, at $\underline{x}_* = (0, 1)$, the first variable is more important than the second. Globally, however, both variables are equally important, because the model is symmetrical.

11.4 Example: Titanic data

Figure 11.2 shows bar plots summarizing the size of oscillations for explanatory variables for the random forest model `titanic_rf` (see Section 4.2.2) for Henry, a 47-year-old man who travelled in the first class (see Section 4.2.5). The longer the bar, the larger the CP-profile oscillations for the particular explanatory variable. The left-hand-side panel presents the variable-importance measures computed by applying estimator $\widehat{vip}_{CP}^{j,uni}(\underline{x}_*)$, given in (11.2), to an equidistant grid of values. The right-hand-side panel shows the results obtained by applying estimator $\widehat{vip}_{CP}^{j,emp}(\underline{x}_*)$, given in (11.3), with an empirical distribution for explanatory variables.

The plots presented in Figure 11.2 indicate that both estimators consistently suggest that the most important variables for the model's prediction for Henry are *gender* and *age*, followed by *class*. However, a remarkable difference can be observed for the *sibsp* variable, which gains in relative importance for estimator $\widehat{vip}_{CP}^{j,uni}(\underline{x}_*)$. In this respect, it is worth recalling that this variable has a very skewed distribution (see Figure 4.3). In particular, a significant mass of the distribution is concentrated at zero, but there have been a few high values observed for the variable. As a result, the of empirical density is very different from a uniform distribution. Hence the difference in the relative importance noted in Figure 11.2.

It is worth noting that, while the variable-importance plot in Figure 11.2 does indicate which explanatory variables are important, it does not describe how do the variables influence the prediction. In that respect, the CP profile for *age* for Henry (see Figure 10.4) suggested that, if Henry were older, this would significantly lower his probability of survival. One the other hand, the CP profile for *sibsp* (see Figure 10.4) indicated that, were Henry not travelling alone, this would increase his chances of survival. Thus, the variable-importance plots should always be accompanied by plots of the relevant CP profiles.

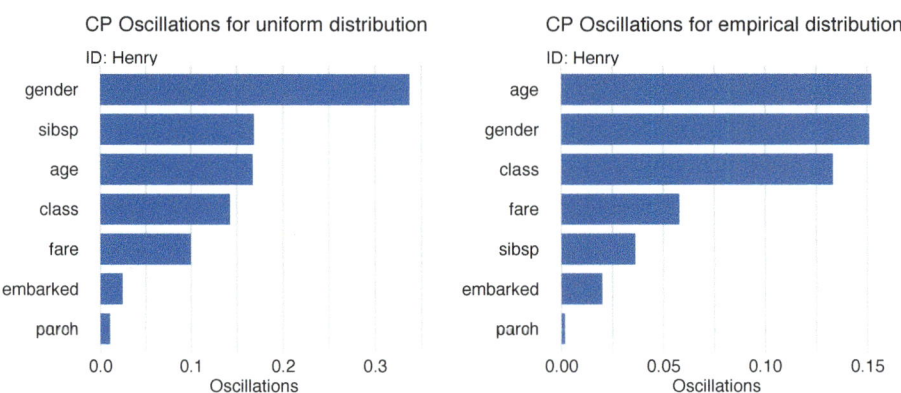

FIGURE 11.2 Variable-importance measures based on ceteris-paribus oscillations estimated by using (left-hand-side panel) a uniform grid of explanatory-variable values and (right-hand-side panel) empirical distribution of explanatory-variables for the random forest model and passenger Henry for the Titanic data.

11.5   Pros and cons

Oscillations of CP profiles are easy to interpret and understand. By using the average of oscillations, it is possible to select the most important variables for an instance prediction. This method can easily be extended to two or more variables.

There are several issues related to the use of the CP oscillations, though. For example, the oscillations may not be of help in situations when the use of CP profiles may itself be problematic (e.g., in the case of correlated explanatory variables or interactions – see Section 10.5). An important issue is that the CP-based variable-importance measures (11.1) do not fulfil the local accuracy condition (see Section 8.2), i.e., they do not sum up to the instance prediction for which they are calculated, unlike the break-down attributions (see Chapter 6) or Shapley values (see Chapter 8).

11.6   Code snippets for R

In this section, we present analysis of CP-profile oscillations as implemented in the **DALEX** package for R. For illustration, we use the random forest model `titanic_rf` (Section 4.2.2). The model was developed to predict the probability of survival after the sinking of the Titanic. Instance-level explanations are calculated for Henry, a 47-year-old male passenger that travelled in the first class (see Section 4.2.5).

We first retrieve the `titanic_rf` model-object and the data frame for Henry via the `archivist` hooks, as listed in Section 4.2.7. We also retrieve the version of the `titanic` data with imputed missing values.

```
titanic_imputed <- archivist::aread("pbiecek/models/27e5c")
titanic_rf <- archivist:: aread("pbiecek/models/4e0fc")
(henry <- archivist::aread("pbiecek/models/a6538"))
```

```
  class gender age sibsp parch fare  embarked
1   1st   male  47     0     0   25 Cherbourg
```

Then we construct the explainer for the model by using the function `explain()` from the **DALEX** package (see Section 4.2.6). We also load the **randomForest** package, as the model was fitted by using function `randomForest()` from this package (see Section 4.2.2) and it is important to have the corresponding `predict()` function available. The model's prediction for Henry is obtained with the help of that function.

```
library("randomForest")
library("DALEX")
explain_rf <- DALEX::explain(model = titanic_rf,
                        data = titanic_imputed[, -9],
                          y = titanic_imputed$survived == "yes",
                      label = "Random Forest")
predict(explain_rf, henry)
```

```
[1] 0.246
```

11.6.1   Basic use of the `predict_parts()` function

To calculate CP-profile oscillations, we use the `predict_parts()` function, already introduced in Section 6.6. In particular, to use estimator $\widehat{vip}_{CP}^{j,uni}(\underline{x}_*)$, defined in (11.2), we specify argument `type="oscillations_uni"`, whereas for estimator $\widehat{vip}_{CP}^{j,emp}(\underline{x}_*)$, defined in (11.3), we specify argument `type="oscillations_emp"`. By default, oscillations are calculated for all explanatory variables. To perform calcualtions only for a subset of variables, one can use the `variables` argument.

In the code below, we apply the function to the explainer-object for the random forest model `titanic_rf` and the data frame for the instance of interest, i.e., `henry`. Additionally, we specify the `type="oscillations_uni"` argument to indicate that we want to compute CP-profile oscillations and the estimated value of the variable-importance measure as defined in (11.2).

```
oscillations_uniform <- predict_parts(explainer = explain_rf,
                           new_observation = henry,
                                    type = "oscillations_uni")
oscillations_uniform
```

```
##      _vname_ _ids_  oscillations
## 2    gender     1   0.33700000
## 4     sibsp     1   0.16859406
## 3       age     1   0.16744554
## 1     class     1   0.14257143
## 6      fare     1   0.09942574
## 7  embarked     1   0.02400000
## 5     parch     1   0.01031683
```

The resulting object is of class `ceteris_paribus_oscillations`, which is a data frame with three variables: `_vname_`, `_ids_`, and `oscillations` that provide, respectively, the name of the variable, the value of the identifier of the instance, and the estimated value of the variable-importance measure. Additionally, the object has also got an overloaded `plot()` function. We can use the latter function to plot the estimated values of the variable-importance measure for the instance of interest. In the code below, before creating the plot, we make the identifier for Henry more explicit. The resulting graph is shown in Figure 11.3.

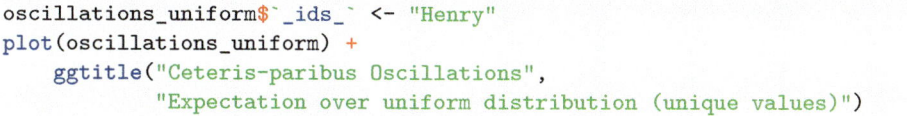

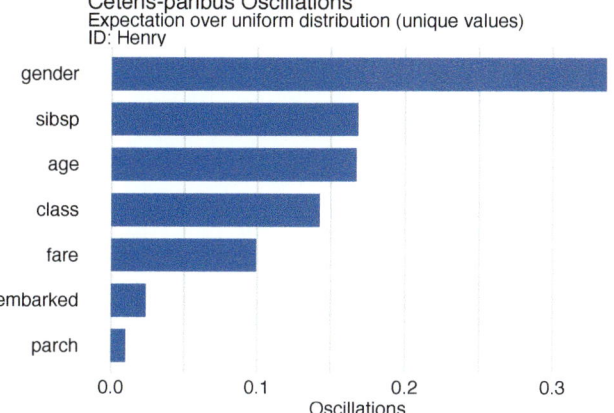

FIGURE 11.3 Variable-importance measures based on ceteris-paribus oscillations estimated by the `oscillations_uni` method of the `predict_parts()` function for the random forest model and passenger Henry for the Titanic data.

11.6.2   Advanced use of the `predict_parts()` function

As mentioned in the previous section, the `predict_parts()` function with argument `type = "oscillations_uni"` computes estimator $\widehat{vip}_{CP}^{\,j,uni}(\underline{x}_*)$, defined in (11.2), while for argument `type="oscillations_emp"` it provides estimator $\widehat{vip}_{CP}^{\,j,emp}(\underline{x}_*)$, defined in (11.3). However, one could also consider applying estimator $\widehat{vip}_{CP}^{\,j,uni}(\underline{x}_*)$ but using a pre-defined grid of values for a continuous explanatory variable. Toward this aim, we can use the `variable_splits` argument to explicitly specify values for the density estimation. Its application is illustrated in the code below for variables *age* and *fare*. Note that, in this case, we use argument `type = "oscillations"`. It is also worth noting that the use of the `variable_splits` argument limits the computations to the variables specified in the argument.

```
oscillations_equidist <- predict_parts(explain_rf, henry,
            variable_splits = list(age = seq(0, 65, 0.1),
                                   fare = seq(0, 200, 0.1),
                                 gender = unique(titanic_imputed$gender),
```

```
                                    class = unique(titanic_imputed$class)),
                    type = "oscillations")
oscillations_equidist
```

```
##    _vname_ _ids_ oscillations
## 3  gender     1    0.3370000
## 1     age     1    0.1677235
## 4   class     1    0.1425714
## 2    fare     1    0.1040790
```

Subsequently, we can use the `plot()` function to construct a bar plot of the estimated values. In the code below, before creating the plot, we make the identifier for Henry more explicit. The resulting graph is shown in Figure 11.4.

```
oscillations_equidist$`_ids_` <- "Henry"
plot(oscillations_equidist) +
    ggtitle("Ceteris-paribus Oscillations",
            "Expectation over specified grid of points")
```

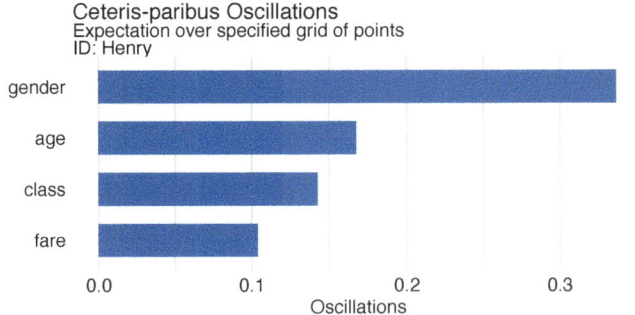

FIGURE 11.4 Variable-importance measures based on ceteris-paribus oscillations estimated by using a specified grid of points for the random forest model and passenger Henry for the Titanic data.

11.7   Code snippets for Python

At this point we are not aware about any Python libraries that would implement the methods presented in the current chapter.

12

Local-diagnostics Plots

12.1 Introduction

It may happen that, despite the fact that the predictive performance of a model is satisfactory overall, the model's predictions for some observations are drastically worse. In such a situation it is often said that "the model does not cover well some areas of the input space".

For example, a model fitted to the data for "typical" patients in a certain hospital may not perform well for patients from another hospital with possible different characteristics. Or, a model developed to evaluate the risk of spring-holiday consumer-loans may not perform well in the case of autumn-loans for Christmas-holiday gifts.

For this reason, in case of important decisions, it is worthwhile to check how does the model behave locally for observations similar to the instance of interest.

In this chapter, we present two local-diagnostics techniques that address this issue. The first one are *local-fidelity plots* that evaluate the local predictive performance of the model around the observation of interest. The second one are *local-stability plots* that assess the (local) stability of predictions around the observation of interest.

12.2 Intuition

Assume that, for the observation of interest, we have identified a set of observations from the training data with similar characteristics. We will call these similar observations "neighbors". The basic idea behind local-fidelity plots is to compare the distribution of residuals (i.e., differences between the observed and predicted value of the dependent variable; see equation (2.1)) for the neighbors with the distribution of residuals for the entire training dataset.

Figure 12.1 presents histograms of residuals for the entire dataset and for
a selected set of 25 neighbors for an instance of interest for the random forest
model for the apartment-prices dataset (Section 4.5.2). The distribution of
residuals for the entire dataset is rather symmetric and centred around 0,
suggesting a reasonable overall performance of the model. On the other hand,
the residuals for the selected neighbors are centred around the value of 500.
This suggests that, for the apartments similar to the one of interest, the model
is biased towards values smaller than the observed ones (residuals are positive,
so, on average, the observed value of the dependent variable is larger than the
predicted value).

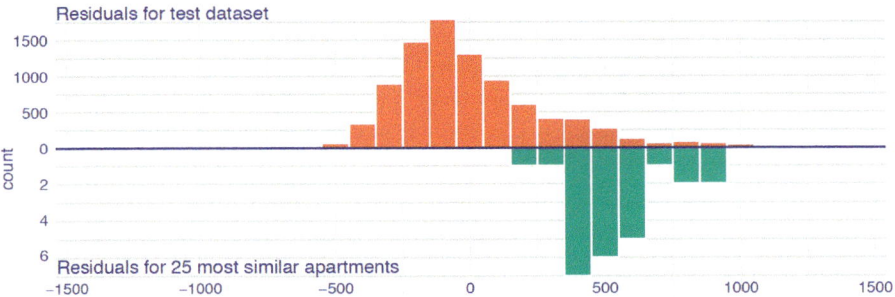

FIGURE 12.1 Histograms of residuals for the random forest model
`apartments_rf` for the apartment-prices dataset. Upper panel: residuals calculated for all observations from the dataset. Bottom panel: residuals calculated
for 25 nearest neighbors of the instance of interest.

The idea behind local-stability plots is to check whether small changes in the
explanatory variables, as represented by the changes within the set of neighbors,
have got much influence on the predictions. Figure 12.2 presents CP profiles
for variable *age* for an instance of interest and its 10 nearest neighbors for the
random forest model for the Titanic dataset (Section 4.2.2). The profiles are
almost parallel and very close to each other. In fact, some of them overlap so
that only 5 different ones are visible. This suggests that the model's predictions
are stable around the instance of interest. Of course, CP profiles for different
explanatory variables may be very different, so a natural question is: which
variables should we examine? The obvious choice is to focus on the variables
that are the most important according to a variable-importance measure such
as the ones discussed in Chapters 6, 8, 9, or 11.

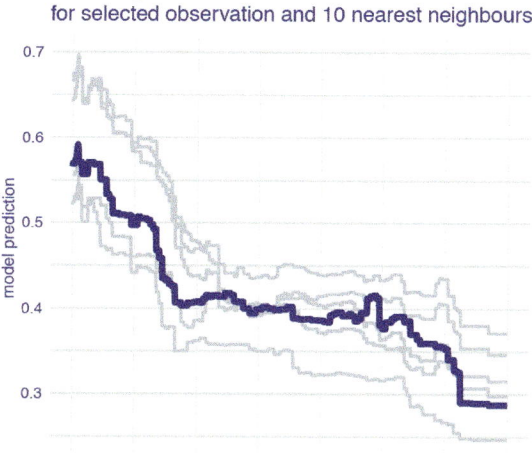

FIGURE 12.2 Ceteris-paribus profiles for a selected instance (dark violet line) and 10 nearest neighbors (light grey lines) for the random forest model for the Titanic data.

12.3 Method

To construct local-fidelity or local-stability plots, we have got to, first, select "neighbors" of the observation of interest. Then, for the fidelity analysis, we have got to calculate and compare residuals for the neighbors. For the stability analysis, we have got to calculate and visualize CP profiles for the neighbors.

In what follows, we discuss each of these steps in more detail.

12.3.1 Nearest neighbors

There are two important questions related to the selection of the neighbors "nearest" to the instance (observation) of interest:

- How many neighbors should we choose?
- What metric should be used to measure the "proximity" of observations?

The answer to both questions is *it depends*.

The smaller the number of neighbors, the more local is the analysis. However, selecting a very small number will lead to a larger variability of the results. In many cases we found that having about 20 neighbors works fine. However, one should always take into account computational time (because a smaller number of neighbors will result in faster calculations) and the size of the dataset (because, for a small dataset, a smaller set of neighbors may be preferred). The metric is very important. The more explanatory variables, the more important is the choice. In particular, the metric should be capable of accommodating variables of different nature (categorical, continuous). Our default choice is the Gower similarity measure:

$$d_{gower}(\underline{x}_i, \underline{x}_j) = \frac{1}{p} \sum_{k=1}^{p} d^k(x_i^k, x_j^k), \tag{12.1}$$

where $\underline{x}_i$ is a p-dimensional vector of values of explanatory variables for the i-th observation and $d^k(x_i^k, x_j^k)$ is the distance between the values of the k-th variable for the i-th and j-th observations. Note that p may be equal to the number of all explanatory variables included in the model, or only a subset of them. Metric $d^k()$ in (12.1) depends on the nature of the variable. For a continuous variable, it is equal to

$$d^k(x_i^k, x_j^k) = \frac{|x_i^k - x_j^k|}{\max(x_1^k, \ldots, x_n^k) - \min(x_1^k, \ldots, x_n^k)},$$

i.e., the absolute difference scaled by the observed range of the variable. On the other hand, for a categorical variable,

$$d^k(x_i^k, x_j^k) = 1_{x_i^k = x_j^k},$$

where 1_A is the indicator function for condition A.

An advantage of the Gower similarity measure is that it can be used for vectors with both categorical and continuous variables. A disadvantage is that it takes into account neither correlation between variables nor variable importance. For a high-dimensional setting, an interesting alternative is the proximity measure used in random forests (Breiman, 2001a), as it takes into account variable importance; however, it requires a fitted random forest model.

Once we have decided on the number of neighbors, we can use the chosen metric to select the required number of observations "closest" to the one of interest.

12.3.2   Local-fidelity plot

Figure 12.1 summarizes two distributions of residuals, i.e., residuals for the neighbors of the observation of interest and residuals for the entire training dataset except for neighbors.

For a typical observation, these two distributions shall be similar. An alarming situation is if the residuals for the neighbors are shifted towards positive or negative values.

Apart from visual examination, we may use statistical tests to compare the two distributions. If we do not want to assume any particular parametric form of the distributions (like, e.g., normal), we may choose non-parametric tests like the Wilcoxon test or the Kolmogorov-Smirnov test. For statistical tests, it is important that the two sets are disjointed.

12.3.3 Local-stability plot

Once neighbors of the observation of interest have been identified, we can graphically compare CP profiles for selected (or all) explanatory variables.

For a model with a large number of variables, we may end up with a large number of plots. In such a case, a better strategy is to focus only on a few most important variables, selected by using a variable-importance measure (see, for example, Chapter 11).

CP profiles are helpful to assess model stability. In addition, we can enhance the plot by adding residuals to them to allow evaluation of the local model-fit. The plot that includes CP profiles for the nearest neighbors and the corresponding residuals is called a local-stability plot.

12.4 Example: Titanic

As an example, we will consider the prediction for Johnny D (see Section 4.2.5) for the random forest model for the Titanic data (see Section 4.2.2).

Figure 12.3 presents a detailed explanation of the elements of a local-stability plot for *age*, a continuous explanatory variable. The plot includes eight nearest neighbors of Johnny D (see Section 4.2.5). The green line shows the CP profile for the instance of interest. Profiles of the nearest neighbors are marked with grey lines. The vertical intervals correspond to residuals; the shorter the interval, the smaller the residual and the more accurate prediction of the model. Blue intervals correspond to positive residuals, red intervals to negative residuals. For an additive model, CP profiles will be approximately parallel. For a model with stable predictions, the profiles should be close to each other. This is not the case of Figure 12.3, in which profiles are quite apart from each other. Thus, the plot suggests potential instability of the model's predictions. Note that there are positive and negative residuals included in the plot. This indicates that, on average, the instance prediction itself should not be biased.

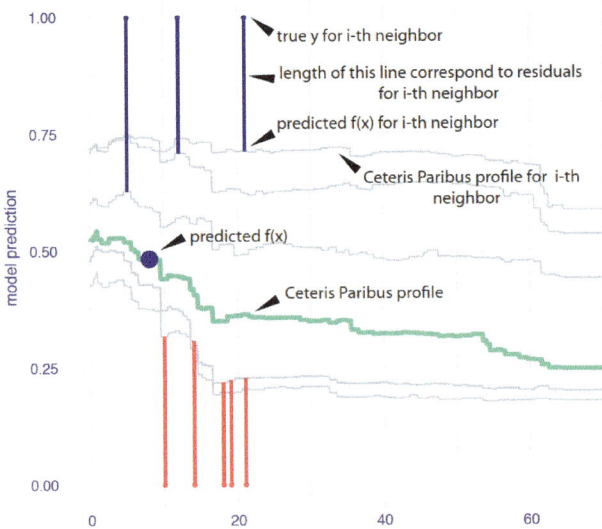

FIGURE 12.3 Elements of a local-stability plot for a continuous explanatory variable. Ceteris-paribus profiles for variable *age* for Johnny D and 5 nearest neighbors for the random forest model for the Titanic data.

12.5   Pros and cons

Local-stability plots may be very helpful to check if the model is locally additive, as for such models the CP profiles should be parallel. Also, the plots can allow assessment whether the model is locally stable, as in that case, the CP profiles should be close to each other. Local-fidelity plots are useful in checking whether the model-fit for the instance of interest is unbiased, as in that case the residuals should be small and their distribution should be symmetric around 0.

The disadvantage of both types of plots is that they are quite complex and lack objective measures of the quality of the model-fit. Thus, they are mainly suitable for exploratory analysis.

12.6   Code snippets for R

In this section, we present local diagnostic plots as implemented in the `DALEX` package for R.

For illustration, we use the random forest model `titanic_rf` (Section 4.2.2). The model was developed to predict the probability of survival after sinking of Titanic. Instance-level explanations are calculated for Henry, a 47-year-old male passenger that travelled in the first class (see Section 4.2.5).

We first retrieve the `titanic_rf` model-object and the data frame for Henry via the `archivist` hooks, as listed in Section 4.2.7. We also retrieve the version of the `titanic` data with imputed missing values.

```
titanic_imputed <- archivist::aread("pbiecek/models/27e5c")
titanic_rf <- archivist:: aread("pbiecek/models/4e0fc")
henry <- archivist::aread("pbiecek/models/a6538")


##   class gender age sibsp parch fare  embarked
## 1   1st   male  47     0     0   25 Cherbourg
```

Then we construct the explainer for the model by using function `explain()` from the `DALEX` package (see Section 4.2.6). We also load the `randomForest` package, as the model was fitted by using function `randomForest()` from this package (see Section 4.2.2) and it is important to have the corresponding `predict()` function available. The model's prediction for Henry is obtained with the help of that function.

```
library("randomForest")
library("DALEX")
explain_rf <- DALEX::explain(model = titanic_rf,
                     data = titanic_imputed[, -9],
                        y = titanic_imputed$survived == "yes",
                    label = "Random Forest")
predict(explain_rf, henry)


## [1] 0.246
```

To construct a local-fidelity plot similar to the one shown Figure 12.1, we can use the `predict_diagnostics()` function from the `DALEX` package. The main arguments of the function are `explainer`, which specifies the name of the explainer-object for the model to be explained, and `new_observation`, which specifies the name of the data frame for the instance for which prediction is of interest. Additional useful arguments are `neighbors`, which specifies the number of observations similar to the instance of interest to be selected (default is 50), and `distance`, the function used to measure the similarity of the observations (by default, the Gower similarity measure is used). Note that function `predict_diagnostics()` has to compute residuals. Thus, we have got to specify the `y` and `residual_function` arguments when using function `explain()` to create the explainer-object (see Section 4.2.6). If the `residual_function` argument is applied with the default `NULL` value, then model residuals are calculated as in (2.1).

In the code below, we perform computations for the random forest model `titanic_rf` and Henry. We select 100 "neighbors" of Henry by using the (default) Gower similarity measure.

```
id_rf <- predict_diagnostics(explainer = explain_rf,
                        new_observation = henry,
                             neighbors = 100)
id_rf
```

```
##
##  Two-sample Kolmogorov-Smirnov test
##
## data:  residuals_other and residuals_sel
## D = 0.47767, p-value = 4.132e-10
## alternative hypothesis: two-sided
```

The resulting object is of class `predict_diagnostics`. It is a list of several components that includes, among others, histograms summarizing the distribution of residuals for the entire training dataset and for the neighbors, as well as the result of the Kolmogorov-Smirnov test comparing the two distributions. The test result is given by default when the object is printed out. In our case, it suggests a statistically significant difference between the two distributions. We can use the `plot()` function to compare the distributions graphically. The resulting graph is shown in Figure 12.4. The plot suggests that the distribution of the residuals for Henry's neighbors might be slightly shifted towards positive values, as compared to the overall distribution.

```
plot(id_rf)
```

Function `predict_diagnostics()` can be also used to construct local-stability plots. Toward this aim, we have got to select the explanatory variable, for which we want to create the plot. We can do it by passing the name of the variable to the `variables` argument of the function. In the code below, we first calculate CP profiles and residuals for *age* and 10 neighbors of Henry.

```
id_rf_age <- predict_diagnostics(explainer = explain_rf,
                       new_observation = henry,
                            neighbors = 10,
                            variables = "age")
```

By applying the `plot()` function to the resulting object, we obtain the local-stability plot shown in Figure 12.5. The profiles are relatively close to each other, suggesting the stability of predictions. There are more negative than

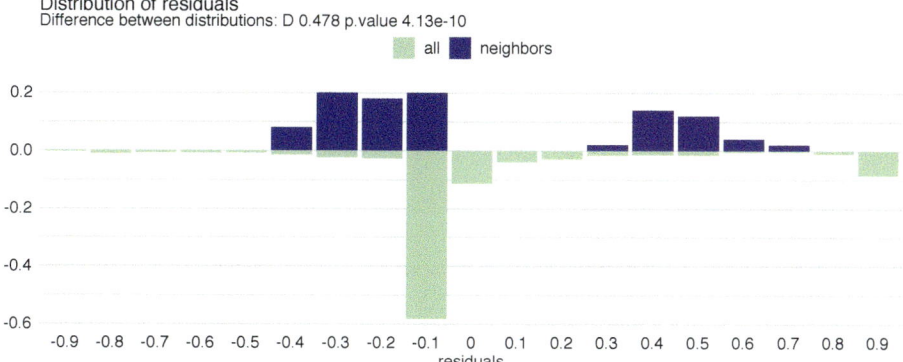

FIGURE 12.4 The local-fidelity plot for the random forest model for the Titanic data and passenger Henry with 100 neighbors.

positive residuals, which may be seen as a signal of a (local) positive bias of the predictions.

```
plot(id_rf_age)
```

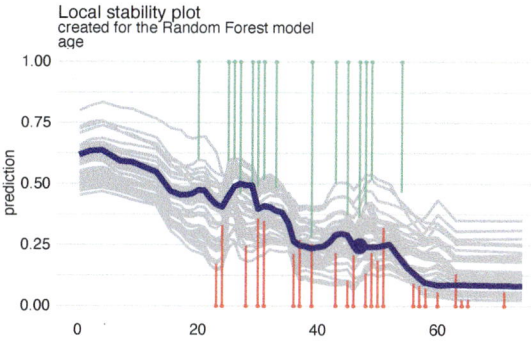

FIGURE 12.5 The local-stability plot for variable *age* in the random forest model for the Titanic data and passenger Henry with 10 neighbors. Note that some profiles overlap, so the graph shows fewer lines.

In the code below, we conduct the necessary calculations for the categorical variable *class* and 10 neighbors of Henry.

```
id_rf_class <- predict_diagnostics(explainer = explain_rf,
                         new_observation = henry,
                                neighbors = 10,
                                variables = "class")
```

By applying the `plot()` function to the resulting object, we obtain the local-stability plot shown in Figure (12.5). The profiles are not parallel, indicating non-additivity of the effect. However, they are relatively close to each other, suggesting the stability of predictions.

```
plot(id_rf_class)
```

12.7   Code snippets for Python

At this point we are not aware of any Python libraries that would implement the methods presented in the current chapter.

13

Summary of Instance-level Exploration

13.1 Introduction

In Part II of the book, we introduced a number of techniques for exploration and explanation of a model's predictions for individual instances. Each chapter was devoted to a single technique. In practice, these techniques are rarely used separately. Rather, it is more informative to combine different insights offered by each technique into a more holistic overview.

Figure 13.1 offers a graphical illustration of the idea. The graph includes results of four different instance-level explanation techniques applied to the random forest model (Section 4.2.2) for the Titanic data (Section 4.1). The instance of interest is Johnny D, an 8-year-old boy who embarked in Southampton and travelled in the first class with no parents nor siblings, and with a ticket costing 72 pounds (Section 4.2.5). Recall that the goal is to predict the probability of survival of Johnny D.

The plots in the first row of Figure 13.1 show results of application of various variable-attribution and variable-importance methods like break-down (BD) plots (Chapter 6), Shapley values (Chapter 8), and local interpretable model-agnostic explanations (LIME, see Chapter 9). The results consistently suggest that the most important explanatory variables, from a point of view of prediction of the probability of survival for Johnny D, are *age*, *gender*, *class*, and *fare*. Note, however, that the picture offered by the additive decompositions may not be entirely correct, because *fare* and *class* are correlated, and there may be an interaction between the effects of *age* and *gender*.

The plots in the second row of Figure 13.1 show ceteris-paribus (CP) profiles (see Chapter 10) for these four most important explanatory variables for Johnny D. The profiles suggest that increasing age or changing the travel class to the second class or to "restaurant staff" would decrease the predicted probability of survival. On the other hand, decreasing the ticket fare, changing gender to female, or changing the travel class to "deck crew" would increase the probability.

The plots in the third row of Figure 13.1 summarize univariate distributions of the four explanatory variables. We see, for instance, that the ticket fare of 72 pounds, which was paid for Johnny D's ticket, was very high and that there were few children among the passengers of Titanic.

Figure 13.1 nicely illustrates that perspectives offered by the different techniques complement each other and, when combined, allow obtaining a more profound insight into the origins of the model's prediction for the instance of interest.

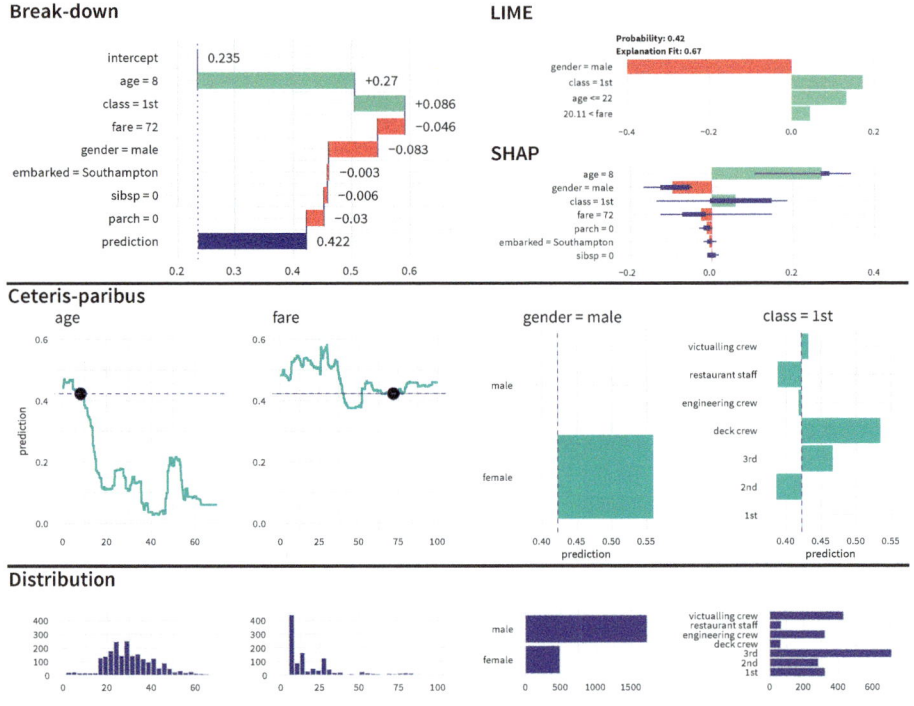

FIGURE 13.1 Results of instance-level-explanation techniques for the random forest model for the Titanic data and passenger Johnny D.

While combining various techniques for instance-level explanation can provide additional insights, it is worth remembering that the techniques are, indeed, different and their suitability may depend on the problem at hand. This is what we discuss in the remainder of the chapter.

13.2   Number of explanatory variables in the model

One of the most important criteria for selection of the exploration and explanation methods is the number of explanatory variables in the model.

13.2.1 Low to medium number of explanatory variables

A low number of variables usually implies that the particular variables have a very concrete meaning and interpretation. An example are variables used in models for the Titanic data presented in Sections 4.2.1 and 4.2.3.

In such a situation, the most detailed information about the influence of the variables on a model's predictions is provided by the CP profiles. In particular, the variables that are most influential for the model's predictions are selected by considering CP-profile oscillations (see Chapter 11) and then illustrated graphically with the help of individual-variable CP profiles (see Chapter 10).

13.2.2 Medium to a large number of explanatory variables

In models with a medium or large number of variables, it is still possible that most (or all) of them are interpretable. An example of such a model is a car-insurance pricing model in which we want to estimate the value of an insurance based on behavioral data that includes 100+ variables about characteristics of the driver and characteristics of the car.

When the number of explanatory variables increases, it becomes harder to show the CP profile for each individual variable. In such situations, the most common approach is to use BD plots, presented in Chapter 6, or plots of Shapley values, discussed in Chapter 8. They allow a quick evaluation whether a particular variable has got a positive or negative effect on a model's prediction; we can also assess the size of the effect. If necessary, it is possible to limit the plots only to the variables with the largest effects.

13.2.3 Very large number of explanatory variables

When the number of explanatory variables is very large, it may be difficult to interpret the role of each single variable. An example of such situations are models for processing of images or texts. In that case, explanatory variables may be individual pixels in image processing or individual characters in text analysis. As such, their individual interpretation is limited. Due to additional issues with computational complexity, it is not feasible to use CP profiles, BD plots, nor Shapley values to evaluate influence of individual values on a model's predictions. Instead, the most common approach is to use LIME, presented in Chapter 9, which works on the context-relevant groups of variables.

13.3   Correlated explanatory variables

When deriving properties for the methods presented in Part II of this book, we often assumed that explanatory variables are independent. Obviously, this is not always the case. For instance, in the case of the data on apartment prices (see Section 4.4.1), the number of rooms and the surface of an apartment will most likely be positively associated. A similar conclusion can be drawn for the travel class and ticket fare for the Titanic data (see Section 4.1.1).

Of course, technically speaking, all the presented methods can be applied also when explanatory variables are correlated. However, in such a case the results may be misleading or unrealistic.

To address the issue, one could consider creating new variables that would be independent. This is sometimes possible using the application-domain knowledge or by using suitable statistical techniques like principal-components analysis. An alternative is to construct two-dimensional CP plots (see Section 10.5) or permute variables in blocks to preserve the correlation structure of variables when computing Shapley values (see Chapter 8) or BD plots (see Chapter 6).

13.4   Models with interactions

In models with interactions, the effect of one explanatory variable may depend on values of other variables. For example, the probability of survival for Titanic passengers may decrease with age, but the effect may be different for different travel-classes.

In such a case, to explore and explain a model's predictions, we have got to consider not individual variables, but sets of variables included in interactions. To identify interactions, we can use iBD plots, as described in Chapter 7. To show effects of interaction, we may use a set of CP profiles. In particular, for the Titanic example, we may use the CP profiles for the *age* variable for instances that differ only in *gender*. The less parallel are such profiles, the larger the effect of interaction.

13.5 Sparse explanations

Predictive models may use hundreds of explanatory variables to yield a prediction for a particular instance. However, for a meaningful interpretation and illustration, most human beings can handle only a very limited (say, less than 10) number of variables. Thus, sparse explanations are of interest. The most common method that is used to construct such explanations is LIME (Chapter 9). However, constructing a sparse explanation for a complex model is not trivial and may be misleading. Hence, care is needed when applying LIME to very complex models.

13.6 Additional uses of model exploration and explanation

In the previous chapters of Part II of the book, we focused on the application of the presented methods to exploration and explanation of predictive models. However, the methods can also be used for other purposes:

- *Model improvement/debugging.* If a model's prediction is particularly bad for a selected observation, then the investigation of the reasons for such a bad performance may provide hints about how to improve the model. In the case of instance predictions, it is easier to detect that a selected explanatory variable should have a different effect than the observed one.

- *Additional domain-specific validation.* Understanding which factors are important for a model's predictions helps in evaluation of the plausibility of the model. If the effects of some explanatory variables on the predictions are observed to be inconsistent with the domain knowledge, this may provide a ground for criticising the model and, eventually, replacing it by another one. On the other hand, if the influence of the variables on the model's predictions is consistent with prior expectations, the user may become more confident with the model. Such confidence is fundamental when the model's predictions are used as a support for taking decisions that may lead to serious consequences, like in the case of, for example, predictive models in medicine.

- *Model selection.* In the case of multiple candidate models, one may use results of the model explanation techniques to select one of the candidates. It is possible that, even if two models are similar in terms of overall performance, one of them may perform much better locally. Consider the following, highly hypothetical example. Assume that a model is sought to predict whether it will rain on a particular day in a region where it rains on half of the days.

Two models are considered: one which simply predicts that it will rain every other day, and another that predicts that it will rain every day since October till March. Arguably, both models are rather unsophisticated (to say the least), but they both predict that, on average, half of the days will be rainy. (We can say that both models are well-calibrated; see Section 15.2.) However, investigation of the instance predictions (for individual days) may lead to a preference for one of them.

- *New knowledge extraction.* Machine-learning models are mainly built for the effectiveness of predictions. As mentioned by Breiman (2001b), it is a different style than the modelling based on the understanding of the phenomena that generated observed values of interests. However, model explanations may sometimes help to extract new and useful knowledge in the field, especially in areas where there is not much domain knowledge yet.

13.7 Comparison of models (champion-challenger analysis)

The techniques for explaining and exploring models have many applications. One of them is the opportunity to compare models.

There are situations when we may be interested in the "champion-challenger" analysis. Let us assume that some institution uses a predictive model, but wants to know if it could get a better model using other modelling techniques. For example, the risk department in a bank may be using logistic regression to assess credit risk. The model may perform satisfactorily and, hence, be considered as the "champion", i.e., the best model in the class of logistic regression models. However, the department may be interested in checking whether a "challenger", i.e., a more complex model developed by using, for instance, boosting or random trees, will not perform better. And if it is performing better, the question of interest is: how does the challenger differ from the champion?

Another reason why we may want to compare models is the fact that the modelling process is iterative itself (see Section 2.2). During the process many versions of models are created, often with different structures, and sometimes with a very similar performance. Comparative analysis allows for better understanding of how these models differ from each other.

Below we present an example of a comparative analysis for the logistic regression model `titanic_lmr` (Section 4.2.1), random forest model `titanic_rf` (Section 4.2.2), boosting model of `titanic_gbm` (Section 4.2.3), and support-vector machine (SVM) model `titanic_svm` (Section 4.2.4). We consider Johnny D (see Section 4.2.5) as the instance of interest.

Note that the models do importantly differ. The random forest and boosting models are tree-based, with a stepped response (prediction) curve. They are complex due to a large number of trees used for prediction. The logistic regression and SVM models lead to continuous and smooth response curves. Their complexity stems from the fact that the logistic regression model includes spline transformations, while the SVM model uses a non-linear kernel function. The differences result in different predicted values of the probability of survival for Johnny D. In particular, the predicted value of the probability is equal to 0.42, 0.77, 0.66, and 0.22 for the random forest, logistic regression, gradient boosting, and SVM model, respectively (see Section 4.2.5).

Figure 13.2 shows the Shapley values (see Chapter 8) for the four models for Johnny D. For the random forest and logistic regression models, similar variables are indicated as important: *age*, *class*, and *gender*. *Class* and *gender* are also important for the gradient boosting model, while for the SVM model, the most important variable is *gender*, followed by *age* and *parch*.

Shapley values for Johnny D

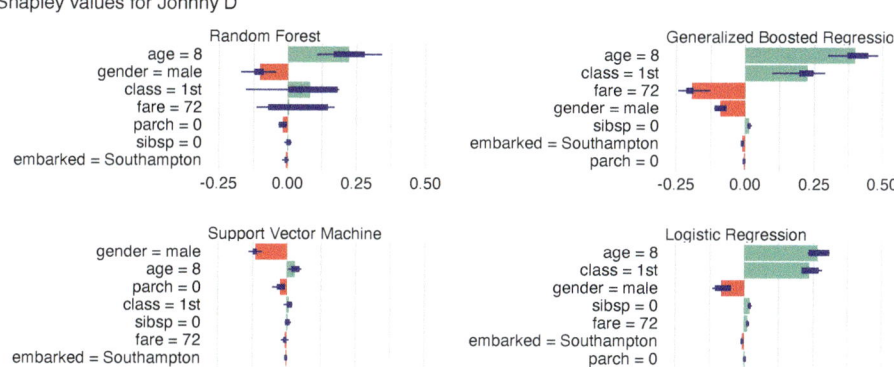

FIGURE 13.2 Shapley values for four different models for the Titanic data and passenger Johnny D.

As mentioned in Chapter 8, Shapley values show additive contributions of explanatory variables to a model's predictions. However, the values may be misleading if there are interactions. In that case, iBD plots, discussed in Chapter 7, might be more appropriate. Figure 13.3 presents the plots for the four models under consideration.

For the SVM model, the most important variable is *gender*, while for the other models the most important variables are *age* and *class*. Remarkably, the iBD plot for the random forest model includes the interaction of *fare* and *class*.

Figure 13.4 shows CP profiles for *age* and *fare* and the four compared models. For *fare*, the logistic regression and SVM models show little effect. A similar conclusion can be drawn for the boosting model, though for this model the

Interaction break-down plots for Johnny D

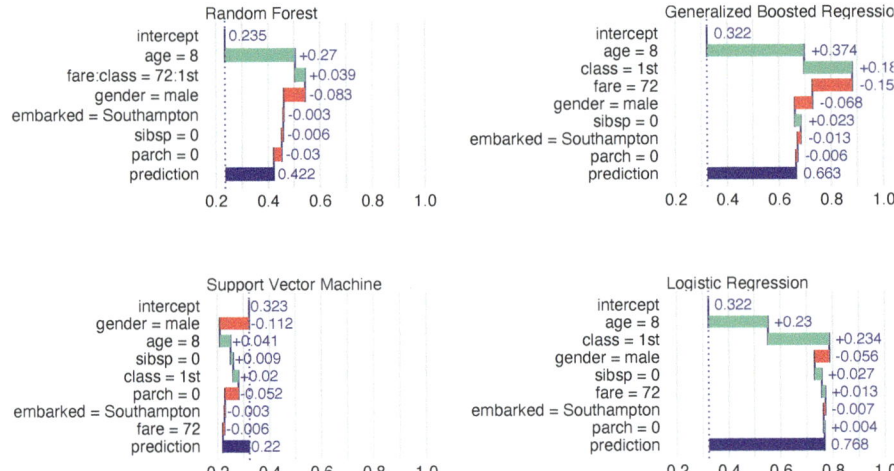

FIGURE 13.3 Interaction break-down plots for four different models for the Titanic data and Johnny D.

profile shows considerable oscillations. The profile for the random forest model suggests a decrease in the predicted probability of survival when the ticket fare increases over about 37 pounds.

For *age*, the CP profile for the SVM model shows, again, considerable oscillations. For all the models, however, the effect of the variable is substantial, with the predicted probability of survival decreasing with increasing age. The effect is most pronounced for the logistic regression model.

Breiman (2001b) described a phenomenon called "Rashomon effect". It means that different models with a similar performance can base their predictions on completely different relations extracted from the same data.

This is the case with the four models presented in this chapter. Although they are all quite effective, for one particular observation, Johnny D, they give different predictions and different explanations.

Despite the differences, the explanations are very useful. For example, the CP profiles for the gradient boosting model exhibit high fluctuations (see Figure 13.4). Such fluctuations suggest overfitting, so one might be inclined not to use this model. The CP profiles for *age*, presented in Figure 13.4, indicate that the SVM model is rather inflexible in that it cannot capture the trend, seen for the remaining three models, that young passengers had a better chance of survival. This could be taken as an argument against the use of the SVM model.

FIGURE 13.4 Ceteris-paribus plots for variables *age* and *fare* for four different models for the Titanic data and passenger Johnny D.

The CP-profiles for the random forest-model are, in general, consistent with the logistic regression model (see Figure 13.4). However, box plots in Figure 13.2 show that explanatory-variable attributions for the random forest model are highly variable. This indicates the presence of interactions in the model. However, for a model with interactions, the additive explanations for individual explanatory variables can be misleading.

Based on this analysis, we might conclude that the most recommendable of the four models is the logistic regression model with splines.

Part III

Dataset Level

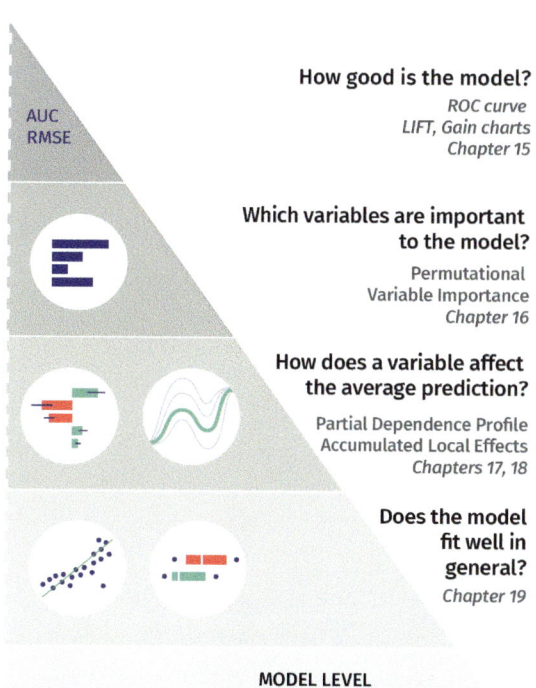

How good is the model?
ROC curve
LIFT, Gain charts
Chapter 15

Which variables are important
to the model?
Permutational
Variable Importance
Chapter 16

How does a variable affect
the average prediction?
Partial Dependence Profile
Accumulated Local Effects
Chapters 17, 18

Does the model
fit well in
general?
Chapter 19

AUC
RMSE

MODEL LEVEL
GLOBAL EXPLANATIONS

14

Introduction to Dataset-level Exploration

In Part II, we focused on instance-level explainers, which help to understand how does a model yield a prediction for a single observation (instance).

In Part III, we concentrate on dataset-level explainers, which help to understand how do the model predictions perform overall, for an entire set of observations? Assuming that the observations form a representative sample from a general population, dataset-level explainers can provide information about the quality of predictions for the population.

The following examples illustrate situations in which dataset-level explainers may be useful:

- We may want to learn which variables are "important" in the model. For instance, we may be interested in predicting the risk of heart attack by using explanatory variables that are derived from the results of medical examinations. If some of the variables do not influence predictions, we could simplify the model by removing the variables.
- We may want to understand how does a selected variable influence the model's predictions? For instance, we may be interested in predicting prices of apartments. Apartment's location is an important factor, but we may want to know which locations lead to higher prices?
- We may want to discover whether there are any observations, for which the model yields wrong predictions. For instance, for a model predicting the probability of survival after a risky treatment, we might want to know whether there are patients for whom the model's predictions are extremely wrong. Identifying such a group of patients might point to, for instance, an incorrect form of an explanatory variable, or even a missed variable.
- We may be interested in an overall "performance" of the model. For instance, we may want to compare two models in terms of the average accuracy of their predictions.

In all cases, measures capturing a particular aspect of a model's performance have to be defined. We will discuss them in subsequent chapters of this part of the book. In particular, in Chapter 15, we discuss measures that are useful for the evaluation of the overall performance of a model. In Chapter 16, we focus on methods that allow evaluation of a variable's effect on a model's predictions. Chapter 17 and Chapter 18 focus on exploration of the effect of selected variables on predictions. Chapter 19 presents an overview of the classical residual-diagnostics tools. Finally, in Chapter 20, we present an overview of all dataset-level explainers introduced in the book.

15

Model-performance Measures

15.1 Introduction

In this chapter, we present measures that are useful for the evaluation of the overall performance of a (predictive) model.

As it was mentioned in Sections 2.1 and 2.5, in general, we can distinguish between the explanatory and predictive approaches to statistical modelling. Breiman (2001b) indicates that validation of a model can be based on evaluation of *goodness-of-fit* (GoF) or on evaluation of predictive accuracy (which we will term *goodness-of-predicton*, GoP). In principle, GoF is mainly used for explanatory models, while GoP is applied for predictive models. In a nutshell, GoF pertains to the question: how well do the model's predictions explain (fit) dependent-variable values of the observations used for developing the model? On the other hand, GoP is related to the question: how well does the model predict the value of the dependent variable for a new observation? For some measures, their interpretation in terms of GoF or GoP depends on whether they are computed by using training or testing data.

The measures may be applied for several purposes, including:

- *model evaluation*: we may want to know how good the model is, i.e., how reliable are the model's predictions (how frequent and how large errors may we expect)?;
- *model comparison*: we may want to compare two or more models in order to choose between them;
- *out-of-sample and out-of-time comparisons*: we may want to check a model's performance when applied to new data to evaluate if performance has not worsened.

Depending on the nature of the dependent variable (continuous, binary, categorical, count, etc.), different model-performance measures may be used. Moreover, the list of useful measures is growing as new applications emerge. In this chapter, we discuss only a selected set of measures, some of which are used in dataset-level exploration techniques introduced in subsequent chapters. We also limit ourselves to the two basic types of dependent variables continuous (including count) and categorical (including binary) considered in our book.

15.2   Intuition

Most model-performance measures are based on the comparison of the model's predictions with the (known) values of the dependent variable in a dataset. For an ideal model, the predictions and the dependent-variable values should be equal. In practice, it is never the case, and we want to quantify the disagreement.

In principle, model-performance measures may be computed for the training dataset, i.e., the data used for developing the model. However, in that case there is a serious risk that the computed values will overestimate the quality of the model's predictive performance. A more meaningful approach is to apply the measures to an independent testing dataset. Alternatively, a bias-correction strategy can be used when applying them to the training data. Toward this aim, various strategies have been proposed, such as cross-validation or bootstrapping (Kuhn and Johnson, 2013; Harrell, 2015; Steyerberg, 2019). In what follows, we mainly consider the simple data-split strategy, i.e., we assume that the available data are split into a training set and a testing set. The model is created on the former, and the latter set is used to assess the model's performance.

It is worth mentioning that there are two important aspects of prediction: *calibration* and *discrimination* (Harrell et al., 1996). Calibration refers to the extent of bias in predicted values, i.e., the mean difference between the predicted and true values. Discrimination refers to the ability of the predictions to distinguish between individual true values. For instance, consider a model to be used for weather forecasts in a region where, on average, it rains half the year. A simple model that predicts that every other day is rainy is well-calibrated because, on average, the resulting predicted risk of a rainy day in a year is 50%, which agrees with the actual situation. However, the model is not very much discriminative (for each calendar day, the probability of a correct prediction is 50%, the same as for a fair-coin toss) and, hence, not very useful.

Thus, in addition to overall measures of GoP, we may need separate measures for calibration and discrimination of a model. Note that, for the latter, we may want to weigh differently the situation when the prediction is, for instance, larger than the true value, as compared to the case when it is smaller. Depending on the decision on how to weigh different types of disagreement, we may need different measures.

In the best possible scenario, we can specify a single model-performance measure before the model is created and then optimize the model for this measure. But, in practice, a more common scenario is to use several performance measures, which are often selected after the model has been created.

15.3   Method

Assume that we have got a training dataset with n observations on p explanatory variables and on a dependent variable Y. Let $\underline{x}_i$ denote the (column) vector of values of the explanatory variables for the i-th observation, and y_i the corresponding value of the dependent variable. We will use $\underline{X} = (x'_1, \ldots, x'_n)$ to denote the matrix of explanatory variables for all n observations, and $\underline{y} = (y_1, \ldots, y_n)'$ to denote the (column) vector of the values of the dependent variable.

The training dataset is used to develop model $f(\hat{\underline{\theta}}; \underline{X})$, where $\hat{\underline{\theta}}$ denotes the estimated values of the model's coefficients. Note that could also use here the "penalized" estimates $\tilde{\underline{\theta}}$ (see Section 2.5). Let $\hat{y}_i$ indicate the model's prediction corresponding to y_i.

The model performance analysis is often based on an independent dataset called a testing set. In some cases, model-performance mesures are based on a leave-one-out approach. We will denote by $\underline{X}_{-i}$ the matrix of explanatory variables when excluding the i-th observation and by $f(\hat{\underline{\theta}}_{-i}; \underline{X}_{-i})$ the model developed for the reduced data. It is worth noting here that the leave-one-out model $f(\hat{\underline{\theta}}_{-i}; \underline{X}_{-i})$ is different from the full-data model $f(\hat{\underline{\theta}}; \underline{X})$. But often they are close to each other and conclusions obtained from one can be transferred to the other. We will use $\hat{y}_{i(-i)}$ to denote the prediction for y_i obtained from model $f(\hat{\underline{\theta}}_{-i}; \underline{X}_{-i})$.

In the subsequent sections, we present various model-performance measures. The measures are applied in essentially the same way if a training or a testing dataset is used. If there is any difference in the interpretation or properties of the measures between the two situations, we will explicitly mention them. Note that, in what follows, we will ignore in the notation the fact that we consider the estimated model $f(\hat{\underline{\theta}}; \underline{X})$ and we will use $f()$ as a generic notation for it.

15.3.1   Continuous dependent variable

15.3.1.1   Goodness-of-fit

The most popular GoF measure for models for a continuous dependent variable is the mean squared-error, defined as

$$MSE(f, \underline{X}, \underline{y}) = \frac{1}{n} \sum_{i}^{n} (\hat{y}_i - y_i)^2 = \frac{1}{n} \sum_{i}^{n} r_i^2, \qquad (15.1)$$

where r_i is the residual for the i-th observation (see also Section 2.3). Thus,

MSE can be seen as a sum of squared residuals. MSE is a convex differentiable function, which is important from an optimization point of view (see Section 2.5). As the measure weighs all differences equally, large residuals have got a high impact on MSE. Thus, the measure is sensitive to outliers. For a "perfect" model, which predicts (fits) all y_i exactly, $MSE = 0$.

Note that MSE is constructed on a different scale from the dependent variable. Thus, a more interpretable variant of this measure is the root-mean-squared-error (RMSE), defined as

$$RMSE(f, \underline{X}, \underline{y}) = \sqrt{MSE(f, \underline{X}, \underline{y})}. \tag{15.2}$$

A popular variant of RMSE is its normalized version, R^2, defined as

$$R^2(f, \underline{X}, \underline{y}) = 1 - \frac{MSE(f, \underline{X}, \underline{y})}{MSE(f_0, \underline{X}, \underline{y})}. \tag{15.3}$$

In (15.3), $f_0()$ denotes a "baseline" model. For instance, in the case of the classical linear regression, $f_0()$ is the model that includes only the intercept, which implies the use of the mean value of Y as a prediction for all observations. R^2 is normalized in the sense that the "perfectly" fitting model leads to $R^2 = 1$, while $R^2 = 0$ means that we are not doing better than the baseline model. In the context of the classical linear regression, R^2 is the familiar coefficient of determination and can be interpreted as the fraction of the total variance of Y "explained" by model $f()$.

Given sensitivity of MSE to outliers, sometimes the median absolute-deviation (MAD) is considered:

$$MAD(f, \underline{X}, \underline{y}) = median(|r_1|, ..., |r_n|). \tag{15.4}$$

MAD is more robust to outliers than MSE. A disadvantage of MAD are its less favourable mathematical properties.

Section 15.4.1 illustrates the use of measures for the linear regression model and the random forest model for the apartment-prices data.

15.3.1.2 Goodness-of-prediction

Assume that a testing dataset is available. In that case, we can use model $f()$, obtained by fitting the model to training data, to predict the values of the dependent variable observed in the testing dataset. Subsequently, we can compute MSE as in (15.1) to obtain the mean squared-prediction-error (MSPE) as a GoP measure (Kutner et al., 2005). By taking the square root of MSPE, we get the root-mean-squared-prediction-error (RMSPE).

In the absence of testing data, one of the most known GoP measures for models for a continuous dependent variable is the predicted sum-of-squares (PRESS), defined as

$$PRESS(f, \underline{X}, \underline{y}) = \sum_{i=1}^{n} (\widehat{y}_{i(-i)} - y_i)^2. \tag{15.5}$$

Thus, PRESS can be seen as a result of the application of the leave-one-out strategy to the evaluation of GoP of a model using the training data. Note that, for the classical linear regression model, there is no need to re-fit the model n times to compute PRESS (Kutner et al., 2005).

Based on PRESS, one can define the predictive squared-error $PSE = PRESS/n$ and the standard deviation error in prediction $SEP = \sqrt{PSE} = \sqrt{PRESS/n}$ (Todeschini, 2010). Another measure gaining in popularity is

$$Q^2(f, \underline{X}, \underline{y}) = 1 - \frac{PRESS(f, \underline{X}, \underline{y})}{\sum_{i=1}^{n} (y_i - \bar{y})^2}. \tag{15.6}$$

It is sometimes called the cross-validated R^2 or the coefficient of prediction (Landram et al., 2005). It appears that $Q^2 \leq R^2$, i.e., the expected accuracy of out-of-sample predictions measured by Q^2 cannot exceed the accuracy of in-sample estimates (Landram et al., 2005). For a "perfect" predictive model, $Q^2 = 1$. It is worth noting that, while R^2 always increases if an explanatory variable is added to a model, Q^2 decreases when "noisy" variables are added to the model (Todeschini, 2010).

The aforementioned measures capture the overall predictive performance of a model. A measure aimed at evaluating discrimination is the *concordance index* (c-index) (Harrell et al., 1996; Brentnall and Cuzick, 2018). It is computed by considering all pairs of observations and computing the fraction of the pairs in which the ordering of the predictions corresponds to the ordering of the true values (Brentnall and Cuzick, 2018). The index assumes the value of 1 in case of perfect discrimination and 0.25 for random discrimination.

Calibration can be assessed by a scatter plot of the predicted values of Y in function of the true ones (Harrell et al., 1996; van Houwelingen, 2000; Steyerberg et al., 2010). The plot can be characterized by its intercept and slope. In case of perfect prediction, the plot should assume the form of a straight line with intercept 0 and slope 1. A deviation of the intercept from 0 indicates overall bias in predictions ("calibration-in-the-large"), while the value of the slope smaller than 1 suggests overfitting of the model (van Houwelingen, 2000; Steyerberg et al., 2010). The estimated values of the coefficients can be used to re-calibrate the model (van Houwelingen, 2000).

15.3.2 Binary dependent variable

To introduce model-performance measures, we, somewhat arbitrarily, label the two possible values of the dependent variable as "success" and "failure". Of course, in a particular application, the meaning of the "success" outcome does not have to be positive nor optimistic; for a diagnostic test, "success" often means detection of disease. We also assume that model prediction $\widehat{y}_i$ takes the form of the predicted probability of success.

15.3.2.1 Goodness-of-fit

If we assign the value of 1 to success and 0 to failure, it is possible to use MSE, RMSE, and MAD, as defined in equations (15.1), (15.2), (15.4), respectively, as a GoF measure. In fact, the MSE obtained in that way is equivalent to the Brier score, which can be also expressed as

$$\sum_{i=1}^{n} \{y_i(1 - \widehat{y}_i)^2 + (1 - y_i)(\widehat{y}_i)^2\}/n.$$

Its minimum value is 0 for a "perfect" model and 0.25 for an "uninformative" model that yields the predicted probability of 0.5 for all observations. The Brier score is often also interpreted as an overall predictive-performance measure for models for a binary dependent variable because it captures both calibration and the concentration of the predictive distribution (Rufibach, 2010).

One of the main issues related to the summary measures based on MSE is that they penalize too mildly for wrong predictions. In fact, the maximum penalty for an individual prediction is equal to 1 (if, for instance, the model yields zero probability for an actual success). To address this issue, the log-likelihood function based on the Bernoulli distribution (see also (2.8)) can be considered:

$$l(f, \underline{X}, \underline{y}) = \sum_{i=1}^{n} \{y_i \ln(\widehat{y}_i) + (1 - y_i)\ln(1 - \widehat{y}_i)\}. \tag{15.7}$$

Note that, in the machine-learning world, function $-l(f, \underline{X}, \underline{y})/n$ is often considered (sometimes also with ln replaced by $\log_2$) and termed "log-loss" or "cross-entropy". The log-likelihood heavily "penalizes" the cases when the model-predicted probability of success $\widehat{y}_i$ is high for an actual failure ($y_i = 0$) and low for an actual success ($y_i = 1$).

The log-likelihood (15.7) can be used to define R^2-like measures (for a review, see, for example, Allison (2014)). One of the variants most often used is the measure proposed by Nagelkerke (1991):

$$R_{bin}^2(f, \underline{X}, \underline{y}) = \frac{1 - \exp\left(\frac{2}{n}\{l(f_0, \underline{X}, \underline{y}) - l(f, \underline{X}, \underline{y})\}\right)}{1 - \exp\left(\frac{2}{n}l(f_0, \underline{X}, \underline{y})\right)}. \tag{15.8}$$

It shares properties of the "classical" R^2, defined in (15.3). In (15.8), $f_0()$ denotes the model that includes only the intercept, which implies the use of the observed fraction of successes as the predicted probability of success. If we denote the fraction by $\hat{p}$, then

$$l(f_0, \underline{X}, \underline{y}) = n\hat{p} \ln \hat{p} + n(1 - \hat{p}) \ln (1 - \hat{p}).$$

15.3.2.2 Goodness-of-prediction

In many situations, consequences of a prediction error depend on the form of the error. For this reason, performance measures based on the (estimated values of) probability of correct/wrong prediction are more often used.

To introduce some of those measures, we assume that, for each observation from the testing dataset, the predicted probability of success $\widehat{y}_i$ is compared to a fixed cut-off threshold, C say. If the probability is larger than C, then we assume that the model predicts success; otherwise, we assume that it predicts failure. As a result of such a procedure, the comparison of the observed and predicted values of the dependent variable for the n observations in a dataset can be summarized in a table similar to Table 15.1.

TABLE 15.1: Confusion table for a classification model with scores $\widehat{y}_i$.

	True value: success	True value: `failure`	Total
$\widehat{y}_i \geq C$, predicted: `success`	True Positive: TP_C	False Positive (Type I error): FP_C	P_C
$\widehat{y}_i < C$, predicted: `failure`	False Negative (Type II error): FN_C	True Negative: TN_C	N_C
Total	S	F	n

In the machine-learning world, Table 15.1 is often referred to as the "confusion table" or "confusion matrix". In statistics, it is often called the "decision table". The counts TP_C and TN_C on the diagonal of the table correspond to the cases when the predicted and observed value of the dependent variable Y coincide. FP_C is the number of cases in which failure is predicted as a success. These are false-positive, or Type I error, cases. On the other hand, FN_C is the count of false-negative, or Type II error, cases, in which success is predicted as failure. Marginally, there are P_C predicted successes and N_C predicted failures, with $P_C + N_C = n$. In the testing dataset, there are S observed successes and F observed failures, with $S + N = n$.

The effectiveness of such a test can be described by various measures. Let us present some of the most popular examples.

The simplest measure of model performance is *accuracy*, defined as

$$ACC_C = \frac{TP_C + TN_C}{n}.$$

It is the fraction of correct predictions in the entire testing dataset. Accuracy is of interest if true positives and true negatives are more important than their false counterparts. However, accuracy may not be very informative when one of the binary categories is much more prevalent (so called unbalanced labels). For example, if the testing data contain 90% of successes, a model that would always predict a success would reach an accuracy of 0.9, although one could argue that this is not a very useful model.

There may be situations when false positives and/or false negatives may be of more concern. In that case, one might want to keep their number low. Hence, other measures, focused on the false results, might be of interest.

In the machine-learning world, two other measures are often considered: *precision* and *recall*. Precision is defined as

$$Precision_C = \frac{TP_C}{TP_C + FP_C} = \frac{TP_C}{P_C}.$$

Precision is also referred to as the *positive predictive value*. It is the fraction of correct predictions among the predicted successes. Precision is high if the number of false positives is low. Thus, it is a useful measure when the penalty for committing the Type I error (false positive) is high. For instance, consider the use of a genetic test in cancer diagnostics, with a positive result of the test taken as an indication of an increased risk of developing a cancer. A false-positive result of a genetic test might mean that a person would have to unnecessarily cope with emotions and, possibly, medical procedures related to the fact of being evaluated as having a high risk of developing a cancer. We might want to avoid this situation more than the false-negative case. The latter would mean that the genetic test gives a negative result for a person that, actually, might be at an increased risk of developing a cancer. However, an increased risk does not mean that the person will develop cancer. And even so, we could hope that we could detect it in due time.

Recall is defined as

$$Recall_C = \frac{TP_C}{TP_C + FN_C} = \frac{TP_C}{S}.$$

Recall is also referred to as *sensitivity* or the *true-positive rate*. It is the fraction of correct predictions among the true successes. Recall is high if the number of false negatives is low. Thus, it is a useful measure when the penalty for committing the Type II error (false negative) is high. For instance, consider the use of an algorithm that predicts whether a bank transaction is fraudulent. A false-negative result means that the algorithm accepts a fraudulent transaction as a legitimate one. Such a decision may have immediate and unpleasant consequences for the bank, because it may imply a non-recoverable loss of money. On the other hand, a false-positive result means that a legitimate transaction is considered as a fraudulent one and is blocked. However, upon further checking, the legitimate nature of the transaction can be confirmed with, perhaps, the annoyed client as the only consequence for the bank.

The harmonic mean of these two measures defines the *F1 score*:

$$F1\ score_C = \frac{2}{\frac{1}{Precision_C} + \frac{1}{Recall_C}} = 2 \cdot \frac{Precision_C \cdot Recall_C}{Precision_C + Recall_C}.$$

F1 score tends to give a low value if either precision or recall is low, and a high value if both precision and recall are high. For instance, if precision is 0, F1 score will also be 0 irrespectively of the value of recall. Thus, it is a useful measure if we have got to seek a balance between precision and recall.

In statistics, and especially in applications in medicine, the popular measures are *sensitivity* and *specificity*. Sensitivity is simply another name for recall. Specificity is defined as

$$Specificity_C = \frac{TN_C}{TN_C + FP_C} = \frac{TN_C}{F}.$$

Specificity is also referred to as the *true-negative rate*. It is the fraction of correct predictions among the true failures. Specificity is high if the number of false positives is low. Thus, as precision, it is a useful measure when the penalty for committing the Type I error (false positive) is high.

The reason why sensitivity and specificity may be more often used outside the machine-learning world is related to the fact that their values do not depend on the proportion S/n (sometimes termed *prevalence*) of true successes. This means that, once estimated in a sample obtained from a population, they may be applied to other populations, in which the prevalence may be different. This is not true for precision, because one can write

$$Precision_C = \frac{Sensitivity_C \cdot \frac{S}{n}}{Sensitivity_C \cdot \frac{S}{n} + Specificity_C \cdot \left(1 - \frac{S}{n}\right)}.$$

All the measures depend on the choice of cut-off C. To assess the form and the strength of dependence, a common approach is to construct the Receiver Operating Characteristic (ROC) curve. The curve plots $Sensitivity_C$ in function of $1 - Specificity_C$ for all possible, ordered values of C. Figure 15.2 presents the ROC curve for the random forest model for the Titanic dataset. Note that the curve indicates an inverse relationship between sensitivity and specificity: by increasing one measure, the other is decreased.

The ROC curve is very informative. For a model that predicts successes and failures at random, the corresponding curve will be equal to the diagonal line. On the other hand, for a model that yields perfect predictions, the ROC curve reduces to two intervals that connect points (0,0), (0,1), and (1,1).

Often, there is a need to summarize the ROC curve with one number, which can be used to compare models. A popular measure that is used toward this aim is the area under the curve (AUC). For a model that predicts successes and failures at random, AUC is the area under the diagonal line, i.e., it is equal to 0.5. For a model that yields perfect predictions, AUC is equal to 1. It appears that, in this case, AUC is equivalent to the c-index (see Section 15.3.1.2).

Another ROC-curve-based measure that is often used is the *Gini coefficient G*. It is closely related to AUC; in fact, it can be calculated as $G = 2 \times AUC - 1$. For a model that predicts successes and failures at random, $G = 0$; for a perfect-prediction model, $G = 1$. Figure 15.2 illustrates the calculation of the Gini coefficient for the random forest model for the Titanic dataset (see Section 4.2.2).

A variant of ROC curve based on precision and recall is a called a precision-recall curve. Figure 15.3 the curve for the random forest model for the Titanic dataset.

The value of the Gini coefficient or, equivalently, of $AUC - 0.5$ allows a comparison of the model-based predictions with random guessing. A measure that explicitly compares a prediction model with a baseline (or null) model is the *lift*. Commonly, random guessing is considered as the baseline model. In that case,

$$Lift_C = \frac{\frac{TP_C}{P_C}}{\frac{S}{n}} = n\frac{Precision_C}{S}.$$

Note that S/n can be seen as the estimated probability of a correct prediction of success for random guessing. On the other hand, TP_C/P_C is the estimated probability of a correct prediction of a success given that the model predicts a success. Hence, informally speaking, the lift indicates how many more (or less) times does the model do better in predicting success as compared to random guessing. As other measures, the lift depends on the choice of cut-off

C. The plot of the lift as a function of P_C is called the *lift chart*. Figure 15.3 presents the lift chart for the random forest model for the Titanic dataset.

Calibration of predictions can be assessed by a scatter plot of the predicted values of Y in function of the true ones. A complicating issue is a fact that the true values are only equal to 0 or 1. Therefore, smoothing techniques or grouping of observations is needed to obtain a meaningful plot (Steyerberg et al., 2010; Steyerberg, 2019).

There are many more measures aimed at measuring the performance of a predictive model for a binary dependent variable. An overview can be found in, e.g., Berrar (2019).

15.3.3 Categorical dependent variable

To introduce model-performance measures for a categorical dependent variable, we assume that $\underline{y}_i$ is now a vector of K elements. Each element y_i^k ($k = 1, \ldots, K$) is a binary variable indicating whether the k-th category was observed for the i-th observation. We assume that, for each observation, only one category can be observed. Thus, all elements of $\underline{y}_i$ are equal to 0 except one that is equal to 1. Furthermore, we assume that a model's prediction takes the form of a vector, $\widehat{\underline{y}}_i$ say, of the predicted probabilities for each of the K categories, with $\widehat{y}_i^k$ denoting the probability for the k-th category. The predicted category is the one with the highest predicted probability.

15.3.3.1 Goodness-of-fit

The log-likelihood function (15.7) can be adapted to the categorical dependent variable case as follows:

$$l(f, \underline{X}, \underline{y}) = \sum_{i=1}^{n} \sum_{k=1}^{K} y_i^k \ln(\widehat{y}_i^k). \tag{15.9}$$

It is essentially the log-likelihood function based on a multinomial distribution. Based on the likelihood, an R^2-like measure can be defined, using an approach similar to the one used in (15.8) for construction of R_{bin}^2 (Harrell, 2015).

15.3.3.2 Goodness-of-prediction

It is possible to extend measures like accuracy, precision, etc., introduced in Section 15.3.2 for a binary dependent variable, to the case of a categorical one. Toward this end, first, a confusion table is created for each category k, treating the category as "success" and all other categories as "failure". Let us denote the counts in the table by TP_k, FP_k, TN_k, and FN_k. Based on the counts, we can compute the average accuracy across all classes as follows:

$$\overline{ACC_C} = \frac{1}{K} \sum_{k=1}^{K} \frac{TP_{C,k} + TN_{C,k}}{n}. \tag{15.10}$$

Similarly, one could compute the average precision, average sensitivity, etc. In the machine-learning world, this approach is often termed "macro-averaging" (Sokolva and Lapalme, 2009; Tsoumakas et al., 2010). The averages computed in that way treat all classes equally.

An alternative approach is to sum the appropriate counts from the confusion tables for all classes, and then form a measure based on the so-computed cumulative counts. For instance, for precision, this would lead to

$$\overline{Precision_{C}}_{\mu} = \frac{\sum_{k=1}^{K} TP_{C,k}}{\sum_{k=1}^{K}(TP_{C,k} + FP_{C,k})}. \tag{15.11}$$

In the machine-learning world, this approach is often termed "micro-averaging" (Sokolva and Lapalme, 2009; Tsoumakas et al., 2010), hence subscript μ for "micro" in (15.11). Note that, for accuracy, this computation still leads to (15.10). The measures computed in that way favour classes with larger numbers of observations.

15.3.4   Count dependent variable

In case of counts, one could consider using MSE or any of the measures for a continuous dependent variable mentioned in Section 15.3.1.1. However, a particular feature of count dependent variables is that their variance depends on the mean value. Consequently, weighing all contributions to MSE equally, as in (15.1), is not appropriate, because the same residual value r_i indicates a larger discrepancy for a smaller count y_i than for a larger one. Therefore, a popular measure of performance of a predictive model for counts is Pearson's statistic:

$$\chi^2(f, \underline{X}, \underline{y}) = \sum_{i=1}^{n} \frac{(\widehat{y}_i - y_i)^2}{\widehat{y}_i} = \sum_{i=1}^{n} \frac{r_i^2}{\widehat{y}_i}. \tag{15.12}$$

From (15.12) it is clear that, if the same residual is obtained for two different observed counts, it is assigned a larger weight for the count for which the predicted value is smaller.

Of course, there are more measures of model performance as well as types of model responses (e.g., censored data). A complete list, even if it could be created, would be beyond the scope of this book.

15.4 Example

15.4.1 Apartment prices

Let us consider the linear regression model `apartments_lm` (see Section 4.5.1) and the random forest model `apartments_rf` (see Section 4.5.2) for the apartment-prices data (see Section 4.4). Recall that, for these data, the dependent variable, the price per square meter, is continuous. Hence, we can use the performance measures presented in Section 15.3.1. In particular, we consider MSE and RMSE.

Figure 15.1 presents a box plot of the absolute values of residuals for the linear regression and random forest models, computed for the testing-data. The computed values of RMSE are also indicated in the plots. The values are very similar for both models; we have already noted that fact in Section 4.5.4.

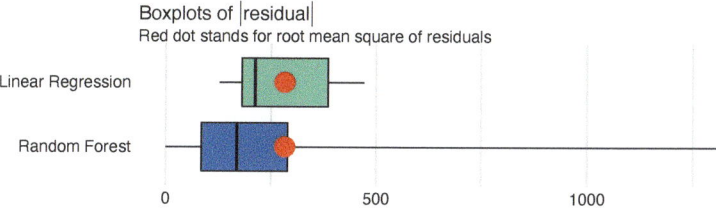

FIGURE 15.1 Box plot for the absolute values of residuals for the linear regression and random forest models for the apartment-prices data. The red dot indicates the RMSE.

In particular, MSE, RMSE, R^2, and MAD values for the linear regression model are equal to 80137, 283.09, 0.901, and 212.7, respectively. For the random forest model, they are equal to 80137, 282.95, 0.901, and 169.1 respectively. The values of the measures suggest that the predicitve performance of the random forest model is slightly better. But is this difference relevant? It should be remembered that development of any random forest model includes a random component. This means that, when a random forest model is fitted to the same dataset several times, but using a different random-number-generation seed, the value of MSE or MAD for the fitted models will fluctuate. Thus, we should consider the values obtained for the linear regression and random forest models for the apartment-prices data as indicating a similar performance of the two models rather than a superiority of one of them.

15.4.2 Titanic data

Let us consider the random forest model `titanic_rf` (see Section 4.2.2) and the logistic regression model `titanic_lmr` (see Section 4.2.1) for the Titanic data (see Section 4.1). Recall that, for these data, the dependent variable is binary, with success defined as survival of the passenger.

First, we take a look at the random forest model. We will illustrate the "confusion table" by using threshold C equal to 0.5, i.e., we will classify passengers as "survivors" and "non-survivors" depending on whether their model-predicted probability of survival was larger than 50% or not, respectively. Table 15.2 presents the resulting table.

TABLE 15.2: Confusion table for the random forest model for the Titanic data. Predicted survival status is equal to *survived* if the model-predicted probability of survival $\hat{y}_i$ is larger than 50%.

	Actual: survived	Actual: died	Total
Predicted: survived	454	60	514
Predicted: died	257	1436	1693
Total	711	1496	2207

Based on the table, we obtain the value of accuracy equal to $(454 + 1436) / 2207 = 0.8564$. The values of precision and recall (sensitivity) are equal to $454/514 = 0.8833$ and $454/711 = 0.6385$, respectively, with the resulting F1 score equal to 0.7412. Specificity is equal to $1436/1496 = 0.9599$.

Figure 15.2 presents the ROC curve for the random forest model. AUC is equal to 0.8595, and the Gini coefficient is equal to 0.719.

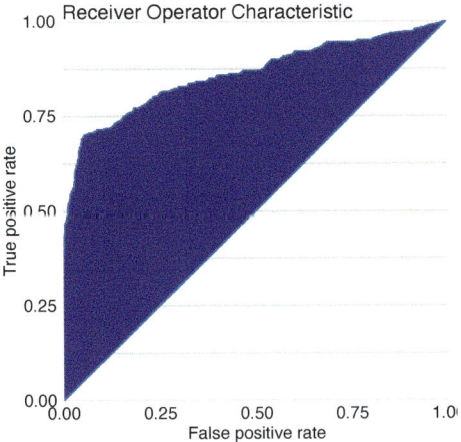

FIGURE 15.2 Receiver Operating Characteristic curve for the random forest model for the Titanic dataset. The Gini coefficient can be calculated as $2\times$ area between the ROC curve and the diagonal (this area is highlighted). The AUC coefficient is defined as an area under the ROC curve.

Figure 15.3 presents the precision-recall curve (left-hand-side panel) and lift chart (right-hand-side panel) for the random forest model.

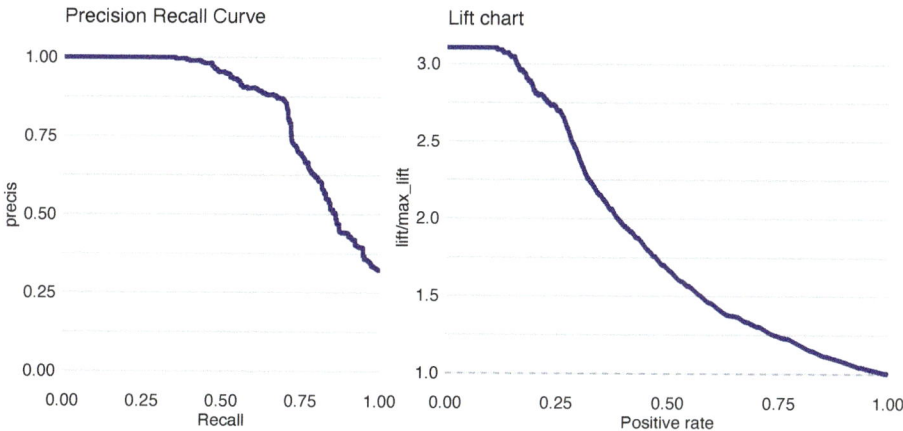

FIGURE 15.3 Precision-recall curve (left panel) and lift chart (right panel) for the random forest model for the Titanic dataset.

Table 15.3 presents the confusion table for the logistic regression model for threshold C equal to 0.5. The resulting values of accuracy, precision, recall (sensitivity), F1 score, and specificity are equal to 0.8043, 0.7522, 0.5851, 0.6582, and 0.9084. The values are smaller than for the random forest model, suggesting a better performance of the latter.

TABLE 15.3: Confusion table for the logisitic regression model for the Titanic data. Predicted survival status is equal to TRUE if the model-predicted probability of survival is larger than 50%.

	Actual: survived	Actual: died	Total
Predicted: survived	416	137	653
Predicted: died	295	1359	1654
Total	711	1496	2207

Left-hand-side panel in Figure 15.4 presents ROC curves for both the logistic regression and the random forest model. The curve for the random forest model lies above the one for the logistic regression model for the majority of the cut-offs C, except for the very high values of the cut-off C. AUC for the logistic regression model is equal to 0.8174 and is smaller than for the random forest model. Right-hand-side panel in Figure 15.4 presents lift charts for both models. Also in this case the curve for the random forest suggests a better performance than for the logistic regression model, except for the very high values of cut-off C.

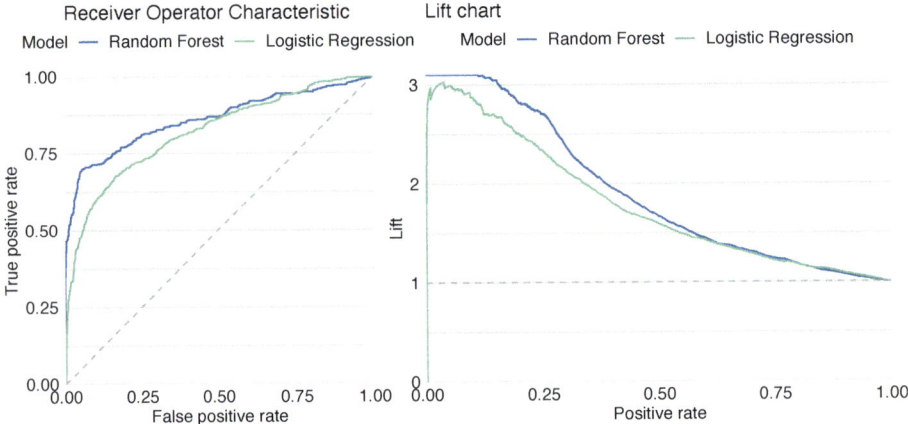

FIGURE 15.4 Receiver Operating Characteristic curves (left panel) and lift charts (right panel) for the random forest and logistic regression models for the Titanic dataset.

15.5   Pros and cons

All model-performance measures presented in this chapter are subject to some limitations. For that reason, many measures are available, as the limitations of a particular measure were addressed by developing an alternative one. For instance, RMSE is frequently used and reported for linear regression models. However, as it is sensitive to outliers, MAD has been proposed as an alternative. In case of predictive models for a binary dependent variable, measures like accuracy, F1 score, sensitivity, and specificity are often considered, depending on the consequences of correct/incorrect predictions in a particular application. However, the value of those measures depends on the cut-off value used for creating predictions. For this reason, ROC curve and AUC have been developed and have become very popular. They are not easily extended to the case of a categorical dependent variable, though.

Given the advantages and disadvantages of various measures and the fact that each may reflect a different aspect of the predictive performance of a model, it is customary to report and compare several of them when evaluating a model's performance.

15.6   Code snippets for R

In this section, we present model-performance measures as implemented in the DALEX package for R. The package covers the most often used measures and methods presented in this chapter. More advanced measures of performance are available in the auditor package for R (Gosiewska and Biecek, 2018). Note that there are also other R packages that offer similar functionality. These include, for instance, packages mlr (Bischl et al., 2016), caret (Kuhn, 2008), tidymodels (Max and Wickham, 2018), and ROCR (Sing et al., 2005).

For illustration purposes, we use the random forest model titanic_rf (see Section 4.2.2) and the logistic regression model titanic_lmr (see Section 4.2.1) for the Titanic data (see Section 4.1). Consequently, the DALEX functions are applied in the context of a binary classification problem. However, the same functions can be used for, for instance, linear regression models.

To illustrate the use of the functions, we first retrieve the titanic_lmr and titanic_rf model-objects via the archivist hooks, as listed in Section 4.2.7. We also retrieve the version of the titanic data with imputed missing values.

```r
titanic_imputed <- archivist::aread("pbiecek/models/27e5c")
titanic_lmr <- archivist::aread("pbiecek/models/58b24")
titanic_rf <- archivist::aread("pbiecek/models/4e0fc")
```

Then we construct the explainers for the models by using function explain() from the DALEX package (see Section 4.2.6). We also load the rms and randomForest packages, as the models were fitted by using functions from those packages and it is important to have the corresponding predict() functions available.

```r
library("rms")
library("DALEX")
explain_lmr <- explain(model = titanic_lmr,
                       data = titanic_imputed[, -9],
                       y = titanic_imputed$survived == "yes",
                       type = "classification",
                       label = "Logistic Regression")

library("randomForest")
explain_rf <- explain(model = titanic_rf,
                      data = titanic_imputed[, -9],
                      y = titanic_imputed$survived == "yes",
                      label = "Random Forest")
```

Function model_performance() calculates, by default, a set of selected model-performance measures. These include MSE, RMSE, R^2, and MAD for linear regression models, and recall, precision, F1, accuracy, and AUC for models

for a binary dependent variable. The function includes the `cutoff` argument
that allows specifying the cut-off value for the measures that require it, i.e.,
recall, precision, F1 score, and accuracy. By default, the cut-off value is set at
0.5. Note that, by default, all measures are computed for the data that are
extracted from the explainer object; these can be training or testing data.

```
(eva_rf <- DALEX::model_performance(explain_rf))
```

```
## Measures for:  classification
## recall      : 0.6385373
## precision   : 0.8832685
## f1          : 0.7412245
## accuracy    : 0.8563661
## auc         : 0.8595467
##
## Residuals:
##      0%      10%      20%      30%      40%      50%      60%      70%
## -0.8920  -0.1140  -0.0240  -0.0080  -0.0040   0.0000   0.0000   0.0100
##      80%      90%     100%
##   0.1400   0.5892   1.0000
```

```
(eva_lr <- DALEX::model_performance(explain_lmr))
```

```
## Measures for:  classification
## recall      : 0.5850914
## precision   : 0.7522604
## f1          : 0.6582278
## accuracy    : 0.8042592
## auc         : 0.81741
##
## Residuals:
##            0%          10%          20%          30%          40%
## -0.98457244  -0.31904861  -0.23408037  -0.20311483  -0.15200813
##            50%          60%          70%          80%          90%
## -0.10318060  -0.06933478   0.05858024   0.29306442   0.73666519
##           100%
##    0.97151255
```

Application of the `DALEX::model_performance()` function returns an object
of class "model_performance", which includes estimated values of several
model-performance measures, as well as a data frame containing the observed
and predicted values of the dependent variable together with their difference,
i.e., residuals. An ROC curve or lift chart can be constructed by applying the
generic `plot()` function to the object. The type of the required plot is indicated
by using argument `geom`. In particular, the argument allows values `geom =
"lift"` for lift charts, `geom = "roc"` for ROC curves, `geom = "histogram"`
for histograms of residuals, and `geom = "boxplot"` for box-and-whisker plots
of residuals. The `plot()` function returns a ggplot2 object. It is possible to
apply the function to more than one object. In that case, the plots for the

models corresponding to each object are combined in one graph. In the code below, we create two `ggplot2` objects: one for a graph containing precision-recall curves for both models, and one for a histogram of residuals. Subsequently, we use the `patchwork` package to combine the graphs in one display.

```r
p1 <- plot(eva_rf, eva_lr, geom = "histogram")
p2 <- plot(eva_rf, eva_lr, geom = "prc")

library("patchwork")
p1 + p2
```

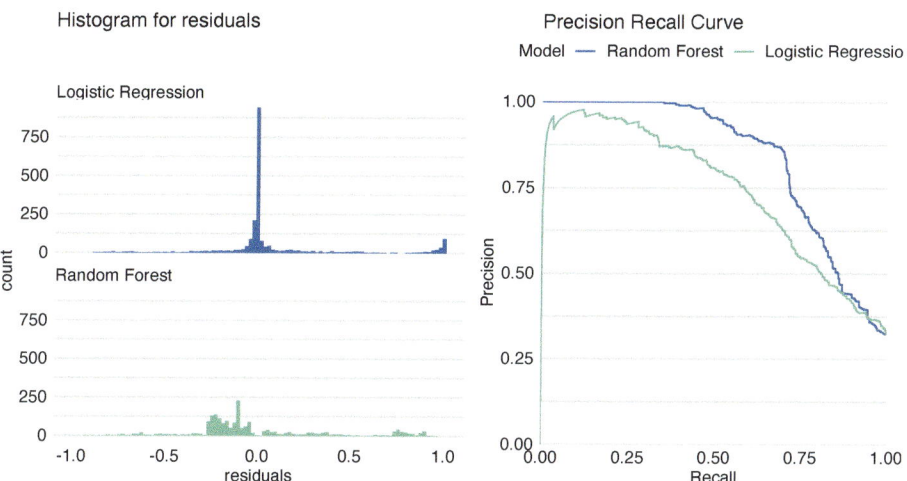

FIGURE 15.5 Precision-recall curves and histograms for residuals obtained by the generic `plot()` function in R for the logistic regression model `titanic_lmr` and the random forest model `titanic_rf` for the Titanic dataset.

The resulting graph is shown in Figure 15.5. Combined with the plot of ROC curves and the lift charts presented in both panels of Figure 15.4, it provides additional insight into the comparison of performance of the two models.

15.7 Code snippets for Python

In this section, we use the `dalex` library for Python. A collection of numerous metrics and performance charts is also available in the popular `sklearn.metrics` library.

For illustration purposes, we use the `titanic_rf` random forest model for the Titanic data developed in Section 4.3.2. Recall that the model is developed to predict the probability of survival for passengers of Titanic.

In the first step, we create an explainer-object that will provide a uniform interface for the predictive model. We use the `Explainer()` constructor for this purpose.

```
import dalex as dx
titanic_rf_exp = dx.Explainer(titanic_rf, X, y,
                 label = "Titanic RF Pipeline")
```

To calculate selected measures of the overall performance, we use the `model_performance()` method. In the syntax below, we apply the `model_type` argument to indicate that we deal with a classification problem, and the `cutoff` argument to specify the cutoff value equal to 0.5. It is worth noting that we get different results than in R. In both cases, the models may differ slightly in implementation and are also trained with a different random seed.

```
mp_rf = titanic_rf_exp.model_performance(model_type = "classification",
          cutoff = 0.5)
mp_rf.result
```

	recall	precision	f1	accuracy	auc
0	0.500703	0.765591	0.605442	0.78976	0.80792

The resulting object can be visualised in many different ways. The code below constructs an ROC curve with AUC measure. Figure 15.6 presents the created plot.

```
import plotly.express as px
from sklearn.metrics import roc_curve, auc
y_score = titanic_rf_exp.predict(X)
fpr, tpr, thresholds = roc_curve(y, y_score)
fig = px.area(x=fpr, y=tpr,
    title=f'ROC Curve (AUC={auc(fpr, tpr):.4f})',
    labels=dict(x='False Positive Rate', y='True Positive Rate'),
    width=700, height=500)
fig.add_shape(
    type='line', line=dict(dash='dash'),
    x0=0, x1=1, y0=0, y1=1)
fig.update_yaxes(scaleanchor="x", scaleratio=1)
fig.update_xaxes(constrain='domain')
fig.show()
```

The code below constructs a plot of FP and TP rates as a function of different thresholds. Figure 15.7 presents the created plot.

```
df = pd.DataFrame({'False Positive Rate': fpr,
        'True Positive Rate': tpr }, index=thresholds)
df.index.name = "Thresholds"
df.columns.name = "Rate"
fig_thresh = px.line(df,
    title='TPR and FPR at every threshold', width=700, height=500)
```

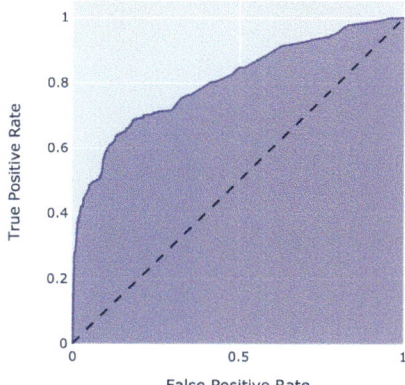

FIGURE 15.6 The ROC curve for the random forest model for the Titanic dataset.

```python
fig_thresh.update_yaxes(scaleanchor="x", scaleratio=1)
fig_thresh.update_xaxes(range=[0, 1], constrain='domain')
fig_thresh.show()
```

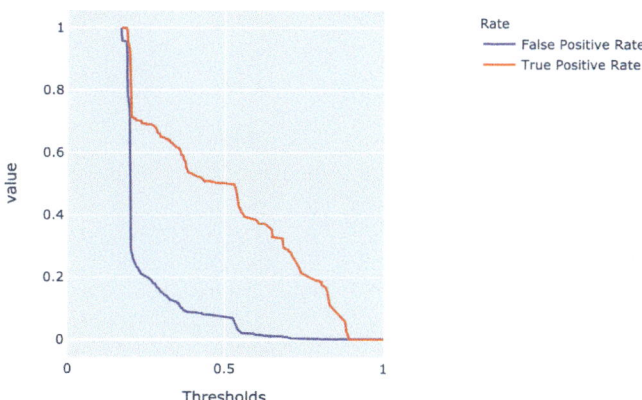

FIGURE 15.7 False-positive and true-positive rates as a function of threshold for the random forest model for the Titanic dataset.

16

Variable-importance Measures

16.1 Introduction

In this chapter, we present a method that is useful for the evaluation of the importance of an explanatory variable. The method may be applied for several purposes.

- *Model simplification*: variables that do not influence a model's predictions may be excluded from the model.
- *Model exploration*: comparison of variables' importance in different models may help in discovering interrelations between the variables. Also, the ordering of variables in the function of their importance is helpful in deciding in which order should we perform further model exploration.
- *Domain-knowledge-based model validation*: identification of the most important variables may be helpful in assessing the validity of the model based on domain knowledge.
- *Knowledge generation*: identification of the most important variables may lead to the discovery of new factors involved in a particular mechanism.

The methods for assessment of variable importance can be divided, in general, into two groups: model-specific and model-agnostic.

For linear models and many other types of models, there are methods of assessing explanatory variable's importance that exploit particular elements of the structure of the model. These are model-specific methods. For instance, for linear models, one can use the value of the normalized regression coefficient or its corresponding p-value as the variable-importance measure. For tree-based ensembles, such a measure may be based on the use of a particular variable in particular trees. A great example in this respect is the variable-importance measure based on out-of-bag data for a random forest model (Breiman, 2001a), but there are also other approaches like methods implemented in the `XgboostExplainer` package (Foster, 2017) for gradient boosting and `randomForestExplainer` (Paluszynska and Biecek, 2017) for random forest.

In this book, we focus on a model-agnostic method that does not assume anything about the model structure. Therefore, it can be applied to any predictive model or ensemble of models. Moreover, and perhaps even more importantly, it allows comparing an explanatory-variable's importance between models with different structures.

16.2   Intuition

We focus on the method described in more detail by Fisher et al. (2019). The main idea is to measure how much does a model's performance change if the effect of a selected explanatory variable, or of a group of variables, is removed? To remove the effect, we use perturbations, like resampling from an empirical distribution or permutation of the values of the variable.

The idea is borrowed from the variable-importance measure proposed by Breiman (2001a) for random forest. If a variable is important, then we expect that, after permuting the values of the variable, the model's performance (as captured by one of the measures discussed in Chapter 15) will worsen. The larger the change in the performance, the more important is the variable.

Despite the simplicity of the idea, the permutation-based approach to measuring an explanatory-variable's importance is a very powerful model-agnostic tool for model exploration. Variable-importance measures obtained in this way may be compared between different models. This property is discussed in detail in Section 16.5.

16.3   Method

Consider a set of n observations for a set of p explanatory variables and dependent variable Y. Let $\underline{X}$ denote the matrix containing, in rows, the (transposed column-vectors of) observed values of the explanatory variables for all observations. Denote by $\underline{y}$ the column vector of the observed values of Y. Let $\hat{\underline{y}} = (f(\underline{x}_1), \ldots, f(\underline{x}_n))'$ denote the corresponding vector of predictions for $\underline{y}$ for model $f()$.

Let $\mathcal{L}(\hat{\underline{y}}, \underline{X}, \underline{y})$ be a loss function that quantifies goodness-of-fit of model $f()$. For instance, $\mathcal{L}()$ may be the value of log-likelihood (see Chapter 15) or any other model performance measure discussed in previous chapter. Consider the following algorithm:

1. Compute $L^0 = \mathcal{L}(\hat{\underline{y}}, \underline{X}, \underline{y})$, i.e., the value of the loss function for the original data. Then, for each explanatory variable X^j included in the model, do steps 2-5.

2. Create matrix $\underline{X}^{*j}$ by permuting the j-th column of $\underline{X}$, i.e., by permuting the vector of observed values of X^j.
3. Compute model predictions $\hat{\underline{y}}^{*j}$ based on the modified data $\underline{X}^{*j}$.
4. Compute the value of the loss function for the modified data:

$$L^{*j} = \mathcal{L}(\hat{\underline{y}}^{*j}, \underline{X}^{*j}, \underline{y}).$$

5. Quantify the importance of X^j by calculating $vip^j_{Diff} = L^{*j} - L^0$ or $vip^j_{Ratio} = L^{*j}/L^0$.

Note that the use of resampling or permuting data in Step 2 involves randomness. Thus, the results of the procedure may depend on the obtained configuration of resampled/permuted values. Hence, it is advisable to repeat the procedure several (many) times. In this way, the uncertainty associated with the calculated variable-importance values can be assessed.

The calculations in Step 5 "normalize" the value of the variable-importance measure with respect to L^0. However, given that L^0 is a constant, the normalization has no effect on the ranking of explanatory variables according to vip^j_{Diff} nor vip^j_{Ratio}. Thus, in practice, often the values of L^{*j} are simply used to quantify a variable's importance.

16.4 Example: Titanic data

In this section, we illustrate the use of the permutation-based variable-importance evaluation by applying it to the random forest model for the Titanic data (see Section 4.2.2). Recall that the goal is to predict survival probability of passengers based on their gender, age, class in which they travelled, ticket fare, the number of persons they travelled with, and the harbour they embarked the ship on.

We use the area under the ROC curve (AUC, see Section 15.3.2.2) as the model-performance measure. Figure 16.1 shows, for each explanatory variable included in the model, the values of $1 - AUC^{*j}$ obtained by the algorithm described in the previous section. Additionally, the plot indicates the value of L^0 by the vertical dashed-line at the left-hand-side of the plot. The lengths of the bars correspond to vip^j_{Diff} and provide the variable-importance measures.

The plot in Figure 16.1 suggests that the most important variable in the model is *gender*. This agrees with the conclusions drawn in the exploratory analysis presented in Section 4.1.1. The next three important variables are *class* (passengers travelling in the first class had a higher chance of survival), *age* (children had a higher chance of survival), and *fare* (owners of more expensive tickets had a higher chance of survival).

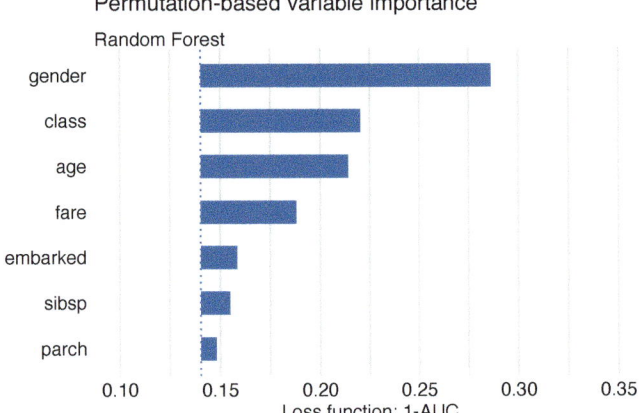

FIGURE 16.1 Single-permutation-based variable-importance measures for the explanatory variables included in the random forest model for the Titanic data using 1-AUC as the loss function.

To take into account the uncertainty related to the use of permutations, we can consider computing the mean values of L^{*j} over a set of, say, 10 permutations. The plot in Figure 16.2 presents the mean values. The only remarkable difference, as compared to Figure 16.1, is the change in the ordering of the *sibsp* and *parch* variables.

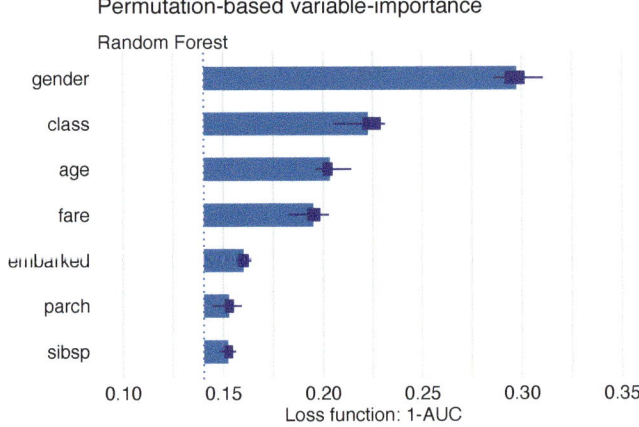

FIGURE 16.2 Means (over 10 permutations) of permutation-based variable-importance measures for the explanatory variables included in the random forest model for the Titanic data using 1-AUC as the loss function.

Plots similar to those presented in Figures 16.1 and 16.2 are useful for comparisons of a variable's importance in different models. Figure 16.3 presents single-permutation results for the random forest, logistic regression (see Section 4.2.1), and gradient boosting (see Section 4.2.3) models. The best result, in terms of the smallest value of L^0, is obtained for the random forest model (as indicated by the location of the dashed lines in the plots). Note that the indicated L^0 value for the model is different from the one indicated in Figure 16.1. This is due to the difference in the set of (random) permutations used to compute the two values.

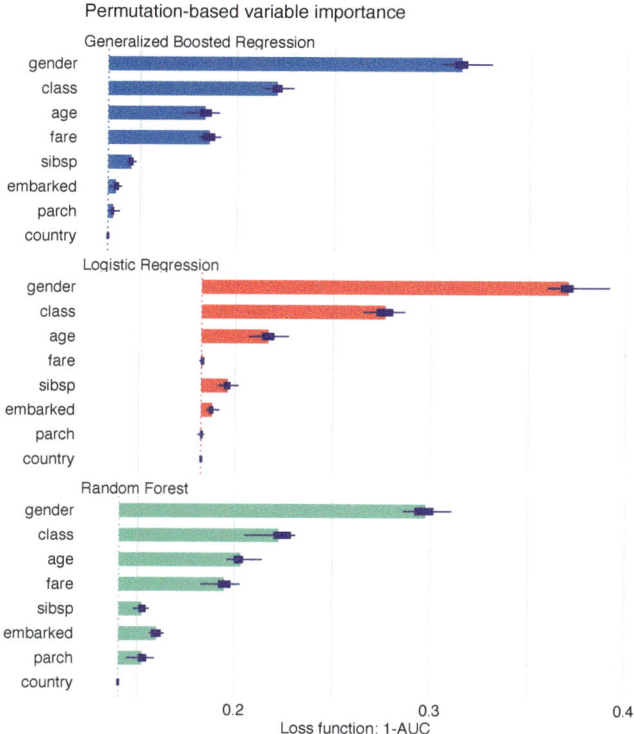

FIGURE 16.3 Single-permutation-based variable-importance measures for the random forest, gradient boosting, and logistic regression models for the Titanic data with 1-AUC as the loss function. Note the different starting locations for the bars, due to differences in the AUC value obtained for the original data for different models.

The plots in Figure 16.3 indicate that *gender* is the most important explanatory variable in all three models, followed by *class* and *age*. Variable *fare*, which is highly correlated with *class*, is important in the random forest and SVM models, but not in the logistic regression model. On the other hand, variable *parch* is, essentially, not important, neither in the gradient boosting nor in the logistic regression model, but it has some importance in the random forest model. *Country* is not important in any of the models. Overall, Figure 16.3

indicates that, in the random forest model, all variables (except of *country*) have got some importance, while in the other two models the effect is mainly limited to *gender*, *class*, and *age* (and *fare* for the gradient boosting model).

16.5   Pros and cons

Permutation-based variable importance offers several advantages. It is a model-agnostic approach to the assessment of the influence of an explanatory variable on a model's performance. The plots of variable-importance measures are easy to understand, as they are compact and present the most important variables in a single graph. The measures can be compared between models and may lead to interesting insights. For example, if variables are correlated, then models like random forest are expected to spread importance across many variables, while in regularized-regression models the effect of one variable may dominate the effect of other correlated variables.

The same approach can be used to measure the importance of a single explanatory variable or a group of variables. The latter is useful for "aspects", i.e., groups of variables that are complementary to each other or are related to a similar concept. For example, in the Titanic example, the *fare* and *class* variables are related to the financial status of a passenger. Instead of assessing the importance of each of these variables separately, we may be interested in their joint importance. Toward this aim, we may compute the permutation-based measure by permuting the values of both variables at the same time.

The main disadvantage of the permutation-based variable-importance measure is its dependence on the random nature of the permutations. As a result, for different permutations, we will, in general, get different results. Also, the value of the measure depends on the choice of the loss function $\mathcal{L}()$. Thus, there is no single, "absolute" measure.

16.6   Code snippets for R

In this section, we present the implementation of the permutation-based variable-importance measure in the `DALEX` package for R. The key function is `model_parts()` that allows computation of the measure. For the purposes of the computation, one can choose among several loss fuctions that include `loss_sum_of_squares()`, `loss_root_mean_square()`, `loss_accuracy()`, `loss_cross_entropy()`, and `loss_one_minus_auc()`. For the definitions of the loss functions, see Chapter 15.

For illustration purposes, we use the random forest model `apartments_rf` for the apartment-prices data (see Section 4.5.2).

We first load the model-object via the `archivist` hook, as listed in Section 4.5.6. We also load the `randomForest` package, as the model was fitted by using function `randomForest()` from this package (see Section 4.5.2) and it is important to have the corresponding `predict()` function available.

Then we construct the explainer for the model by using the function `explain()` from the `DALEX` package (see Section 4.2.6). Note that we use the `apartments_test` data frame without the first column, i.e., the *m2.price* variable, in the `data` argument. This will be the dataset to which the model will be applied (see Section 4.5.5). The *m2.price* variable is explicitly specified as the dependent variable in the `y` argument.

```r
library("DALEX")
library("randomForest")
apartments_rf <- archivist::aread("pbiecek/models/fe7a5")
explainer_rf <- DALEX::explain(model = apartments_rf,
                               data = apartments_test[,-1],
                               y = apartments_test$m2.price,
                               label = "Random Forest")
```

A popular loss function is the root-mean-square-error (RMSE) function (15.2). It is implemented in the `DALEX` package as the `loss_root_mean_square()` function. The latter requires two arguments: `observed`, which indicates the vector of observed values of the dependent variable, and `predicted`, which specifies the object (either vector or a matrix, as returned from the model-specific `predict()` function) with the predicted values. The original-testing-data value L^0 of RMSE for the random forest model can be obtained by applying the `loss_root_mean_square()` in the form given below.

```r
loss_root_mean_square(observed = apartments_test$m2.price,
                      predicted = predict(apartments_rf, apartments_test))
```

```
## [1] 282.9519
```

To compute the permutation-based variable-importance measure, we apply the `model_parts()` function. Note that it is a wrapper for function `feature_importance()` from the `ingredients` package. The only required argument is `explainer`, which indicates the explainer-object (obtained with the help of the `explain()` function, see Section 4.2.6) for the model to be explained. The other (optional) arguments are:

- `loss_function`, the loss function to be used (by default, it is the `loss_root_mean_square` function).
- `type`, the form of the variable-importance measure, with values `"raw"` resulting in the computation of $\mathcal{L}()$, `"difference"` yielding vip_{Diff}^{j}, and `"ratio"` providing vip_{Ratio}^{j} (see Section 16.3).

- `variables`, a character vector providing the names of the explanatory variables, for which the variable-importance measure is to be computed. By default, `variables = NULL`, in which case computations are performed for all variables in the dataset.
- `variable_groups`, a list of character vectors of names of explanatory variables. For each vector, a single variable-importance measure is computed for the joint effect of the variables which names are provided in the vector. By default, `variable_groups = NULL`, in which case variable-importance measures are computed separately for all variables indicated in the `variables` argument.
- `B`, the number of permutations to be used for the purpose of calculation of the (mean) variable-importance measures, with `B = 10` used by default. To get a single-permutation-based measure, use `B = 1`.
- `N`, the number of observations that are to be sampled from the data available in the explainer-object for the purpose of calculation of the variable-importance measure; by default, `N = 1000` is used; if `N = NULL`, the entire dataset is used.

To compute a single-permutation-based value of the RMSE for all the explanatory variables included in the random forest model `apartments_rf`, we apply the `model_parts()` function to the model's explainer-object as shown below. We use the `set.seed()` function to make the process of random selection of the permutation repeatable.

```
set.seed(1980)
model_parts(explainer = explainer_rf,
         loss_function = loss_root_mean_square,
                    B = 1)
```

```
##                  variable mean_dropout_loss          label
## 1            _full_model_          271.9089 Random Forest
## 2 construction.year          389.4840 Random Forest
## 3             no.rooms          396.0281 Random Forest
## 4                floor          436.6190 Random Forest
## 5              surface          462.7374 Random Forest
## 6             district          794.7619 Random Forest
## 7            _baseline_         1095.4724 Random Forest
```

Note that the outcome is identical to the following call below (results not shown).

```
set.seed(1980)
model_parts(explainer = explainer_rf,
         loss_function = loss_root_mean_square,
                    B = 1,
            variables = colnames(explainer_rf$data))
```

However, if we use a different ordering of the variables in the `variables` argument, the result is slightly different:

```
set.seed(1980)
vars <- c("surface","floor","construction.year","no.rooms","district")
model_parts(explainer = explainer_rf,
        loss_function = loss_root_mean_square,
                    B = 1,
            variables = vars)
```

```
##                variable mean_dropout_loss          label
## 1           _full_model_          271.9089 Random Forest
## 2 construction.year          393.1586 Random Forest
## 3               no.rooms          396.0281 Random Forest
## 4                  floor          440.9293 Random Forest
## 5                surface          483.1104 Random Forest
## 6               district          794.7619 Random Forest
## 7            _baseline_         1095.4724 Random Forest
```

This is due to the fact that, despite the same seed, the first permutation is now selected for the *surface* variable, while in the previous code the same permutation was applied to the values of the *floor* variable.

To compute the mean variable-importance measure based on 50 permutations and using the RMSE difference vip^j_{Diff} (see Section 16.3), we have got to specify the appropriate values of the B and `type` arguments.

```
set.seed(1980)
(vip.50 <- model_parts(explainer = explainer_rf,
                    loss_function = loss_root_mean_square,
                                B = 50,
                             type = "difference"))
```

```
##                variable mean_dropout_loss          label
## 1           _full_model_            0.0000 Random Forest
## 2               no.rooms          117.4678 Random Forest
## 3 construction.year          122.4445 Random Forest
## 4                  floor          162.4554 Random Forest
## 5                surface          182.4368 Random Forest
## 6               district          563.7343 Random Forest
## 7            _baseline_          843.0472 Random Forest
```

To obtain a graphical illustration, we apply the `plot()` function to the `vip.50` object.

```
library("ggplot2")
plot(vip.50) +
  ggtitle("Mean variable-importance over 50 permutations", "")
```

The resulting graph is presented in Figure 16.4. The bars in the plot indicate the mean values of the variable-importance measures for all explanatory variables. Box plots are added to the bars to provide an idea about the distribution of the values of the measure across the permutations.

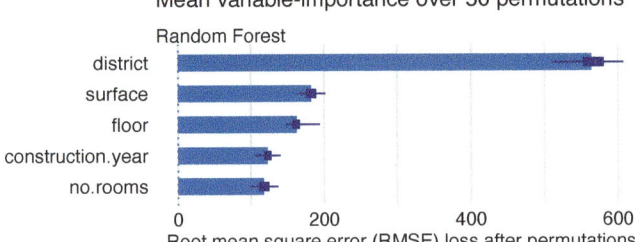

FIGURE 16.4 Mean variable-importance calculated by using 50 permutations and the root-mean-squared-error loss-function for the random forest model `apartments_rf` for the apartment-prices data. Plot obtained by using the generic `plot()` function in R.

Variable-importance measures are a very useful tool for model comparison. We will illustrate this application by considering the random forest model, linear-regression model (Section 4.5.1), and support-vector-machine (SVM) model (Section 4.5.3) for the apartment prices dataset. The models differ in their flexibility and structure; hence, it may be of interest to compare them.

We first load the necessary model-objects via the `archivist` hooks, as listed in Section 4.5.6.

```
apartments_lm  <- archivist::aread("pbiecek/models/55f19")
apartments_svm <- archivist::aread("pbiecek/models/d2ca0")
```

Then we construct the corresponding explainer-objects. We also load the `e1071` package, as it is important to have a suitable `predict()` function available for the SVM model.

```
explainer_lm <- DALEX::explain(model = apartments_lm,
                               data = apartments_test[,-1],
                               y = apartments_test$m2.price,
                               label = "Linear Regression")

library("e1071")
explainer_svm <- DALEX::explain(model = apartments_svm,
                                data = apartments_test[,-1],
                                y = apartments_test$m2.price,
                                label = "Support Vector Machine")
```

Subsequently, we compute mean values of the permutation-based variable-importance measure for 50 permutations and the RMSE loss function. Note that we use the `set.seed()` function to make the process of random selection of the permutation repeatable. By specifying `N = NULL` we include all the data from the apartments dataset in the calculations.

```
vip_lm  <- model_parts(explainer = explainer_lm,  B = 50, N = NULL)
vip_rf  <- model_parts(explainer = explainer_rf,  B = 50, N = NULL)
vip_svm <- model_parts(explainer = explainer_svm, B = 50, N = NULL)
```

Finally, we apply the `plot()` function to the created objects to obtain a single plot with the variable-importance measures for all three models.

```
library("ggplot2")
plot(vip_rf, vip_svm, vip_lm) +
  ggtitle("Mean variable-importance over 50 permutations", "")
```

The resulting graph is presented in Figure 16.5. The plots suggest that the best result, in terms of the smallest value of L^0, is obtained for the SVM model (as indicated by the location of the dashed lines in the plots). The length of bars indicates that *district* is the most important explanatory variable in all three models, followed by *surface* and *floor*. *Construction year* is the fourth most important variable for the random forest and SVM models, but it is not important in the linear-regression model at all. We will investigate the reason for this difference in the next chapter.

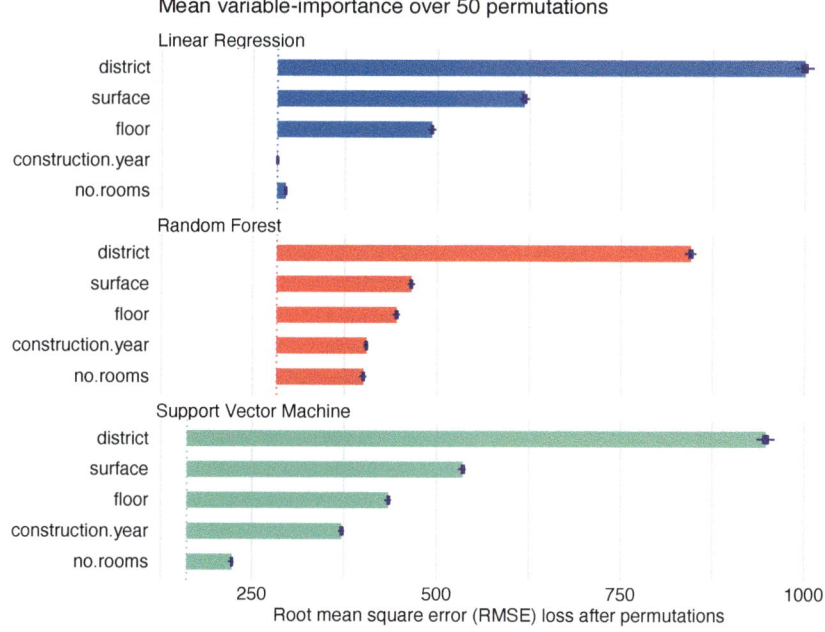

FIGURE 16.5 Mean variable-importance calculated using 50 permutations and the root-mean-squared-error loss for the random forest, support-vector-machine, and linear-regression models for the apartment-prices data.

16.7   Code snippets for Python

In this section, we use the `dalex` library for Python. The package covers all methods presented in this chapter. It is available on `pip` and `GitHub`.

For illustration purposes, we use the `titanic_rf` random forest model for the Titanic data developed in Section 4.3.2. Recall that the model is developed to predict the probability of survival for passengers of Titanic.

In the first step, we create an explainer-object that will provide a uniform interface for the predictive model. We use the `Explainer()` constructor for this purpose.

```
import dalex as dx
titanic_rf_exp = dx.Explainer(titanic_rf, X, y,
               label = "Titanic RF Pipeline")
```

To calculate the variable-importance measure, we use the `model_parts()` method. By default it performs B = 10 permutations of variable importance calculated on N = 1000 observations.

```
mp_rf = titanic_rf_exp.model_parts()
mp_rf.result
```

	variable	dropout_loss	label
0	_full_model_	0.150266	Titanic RF Pipeline
1	parch	0.152570	Titanic RF Pipeline
2	sibsp	0.153574	Titanic RF Pipeline
3	embarked	0.156332	Titanic RF Pipeline
4	fare	0.160662	Titanic RF Pipeline
5	age	0.162763	Titanic RF Pipeline
6	class	0.185076	Titanic RF Pipeline
7	gender	0.377193	Titanic RF Pipeline
8	_baseline_	0.500706	Titanic RF Pipeline

The obtained results can be visualised by using the `plot()` method. Results are presented in Figure 16.6.

```
mp_rf.plot()
```

The `model_parts()` method in Python allows similar arguments as the corresponding function in the `DALEX` package in R (see Section 16.6). These include, for example, the `loss_function` argument (with values like, e.g. ,`'rmse'`

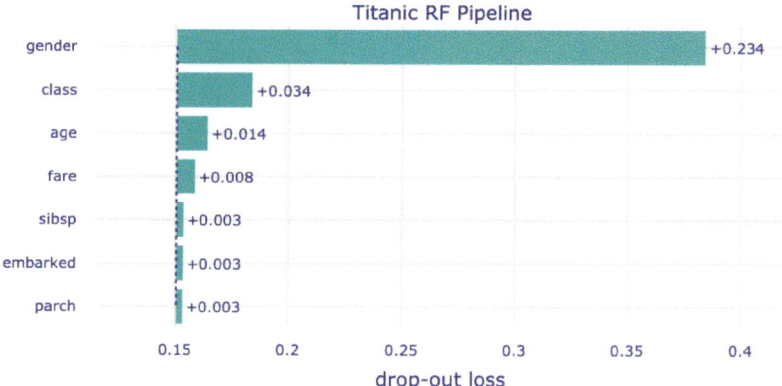

FIGURE 16.6 Mean variable-importance calculated by using 10 permutations and the root-mean-squared-error loss-function for the random forest model for the Titanic data.

or `'1-auc'`); the `type` argument, (with values `'variable_importance'`, `'ratio'`, `'difference'`); and the `variable_groups` argument that allows specifying groups of explanatory variables, for which a single variable-importance measure should be computed.

In the code below, we illustrate the use of the `variable_groups` argument to specify two groups of variables. The resulting plot is presented in Figure 16.7.

```
vi_grouped = titanic_rf_exp.model_parts(
            variable_groups={'personal': ['gender', 'age',
                                          'sibsp', 'parch'],
                             'wealth': ['class', 'fare']})
vi_grouped.result
```

	variable	dropout_loss	label
0	_full_model_	0.150312	Titanic RF Pipeline
1	wealth	0.206971	Titanic RF Pipeline
2	personal	0.398908	Titanic RF Pipeline
3	_baseline_	0.497763	Titanic RF Pipeline

```
vi_grouped.plot()
```

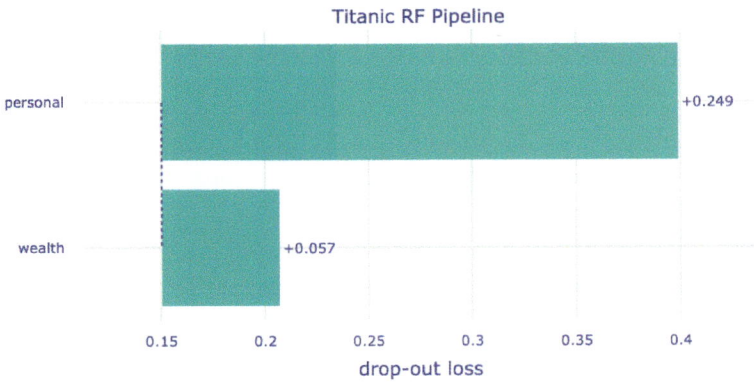

FIGURE 16.7 Mean variable-importance calculated for two groups of variables for the random forest model for the Titanic data.

17

Partial-dependence Profiles

17.1 Introduction

In this chapter, we focus on partial-dependence (PD) plots, sometimes also called PD profiles. They were introduced in the context of gradient boosting machines (GBM) by Friedman (2000). For many years, PD profiles went unnoticed in the shadow of GBM. However, in recent years, they have become very popular and are available in many data-science-oriented packages like **DALEX** (Biecek, 2018), **iml** (Molnar et al., 2018), **pdp** (Greenwell, 2017) or **PDPbox** (Jiangchun, 2018).

The general idea underlying the construction of PD profiles is to show how does the expected value of model prediction behave as a function of a selected explanatory variable? For a single model, one can construct an overall PD profile by using all observations from a dataset, or several profiles for sub-groups of the observations. Comparison of sub-group-specific profiles may provide important insight into, for instance, the stability of the model's predictions.

PD profiles are also useful for comparisons of different models:

- *Agreement between profiles for different models is reassuring.* Some models are more flexible than others. If PD profiles for models, which differ with respect to flexibility, are similar, we can treat it as a piece of evidence that the more flexible model is not overfitting and that the models capture the same relationship.
- *Disagreement between profiles may suggest a way to improve a model.* If a PD profile of a simpler, more interpretable model disagrees with a profile of a flexible model, this may suggest a variable transformation that can be used to improve the interpretable model. For example, if a random forest model indicates a non-linear relationship between the dependent variable and an explanatory variable, then a suitable transformation of the explanatory variable may improve the fit or performance of a linear-regression model.
- *Evaluation of model performance at boundaries.* Models are known to have different behaviour at the boundaries of the possible range of a dependent variable, i.e., for the largest or the lowest values. For instance, random forest models are known to shrink predictions towards the average, whereas support-vector machines are known for a larger variance at edges. Comparison of

PD profiles may help to understand the differences in models' behaviour at boundaries.

17.2   Intuition

To show how does the expected value of model prediction behave as a function of a selected explanatory variable, the average of a set of individual ceteris-paribus (CP) profiles can be used. Recall that a CP profile (see Chapter 10) shows the dependence of an instance-level prediction on an explanatory variable. A PD profile is estimated by the mean of the CP profiles for all instances (observations) from a dataset.

Note that, for additive models, CP profiles are parallel. In particular, they have got the same shape. Consequently, the mean retains the shape, while offering a more precise estimate. However, for models that, for instance, include interactions, CP profiles may not be parallel. In that case, the mean may not necessarily correspond to the shape of any particular profile. Nevertheless, it can still offer a summary of how (in general) do the model's predictions depend on changes in a given explanatory variable.

The left-hand-side panel of Figure 17.1 presents CP profiles for the explanatory variable *age* in the random forest model `titanic_rf` (see Section 4.2.2) for 25 randomly selected instances (observations) from the Titanic dataset (see Section 4.1). Note that the profiles are not parallel, indicating non-additive effects of explanatory variables. The right-hand-side panel shows the mean of the CP profiles, which offers an estimate of the PD profile. Clearly, the shape of the PD profile does not capture, for instance, the shape of the group of five CP profiles shown at the top of the panel. Nevertheless, it does seem to reflect the fact that the majority of CP profiles suggest a substantial drop in the predicted probability of survival for the ages between 2 and 18.

17.3   Method

17.3.1   Partial-dependence profiles

The value of a PD profile for model $f()$ and explanatory variable X^j at z is defined as follows:

$$g_{PD}^j(z) = E_{\underline{X}^{-j}}\{f(X^{j|=z})\}. \tag{17.1}$$

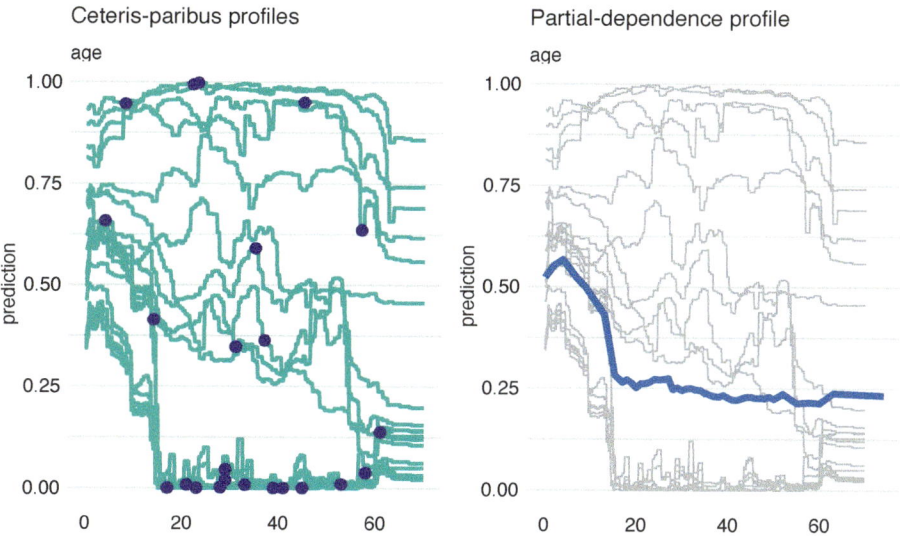

FIGURE 17.1 Ceteris-paribus (CP) and partial-dependence (PD) profiles for the random forest model for 25 randomly selected observations from the Titanic dataset. Left-hand-side plot: CP profiles for *age*; blue dots indicate the age and the corresponding prediction for the selected observations. Right-hand-side plot: CP profiles (grey lines) and the corresponding PD profile (blue line).

Thus, it is the expected value of the model predictions when X^j is fixed at z over the (marginal) distribution of $\underline{X}^{-j}$, i.e., over the joint distribution of all explanatory variables other than X^j. Or, in other words, it is the expected value of the CP profile for X^j, defined in (10.1), over the distribution of $\underline{X}^{-j}$.

Usually, we do not know the true distribution of $\underline{X}^{-j}$. We can estimate it, however, by the empirical distribution of n, say, observations available in a training dataset. This leads to the use of the mean of CP profiles for X^j as an estimator of the PD profile:

$$\hat{g}^j_{PD}(z) = \frac{1}{n} \sum_{i=1}^n f(\underline{x}_i^{j|=z}). \tag{17.2}$$

17.3.2 Clustered partial-dependence profiles

As it has been already mentioned, the mean of CP profiles is a good summary if the profiles are parallel. If they are not parallel, the average may not adequately represent the shape of a subset of profiles. To deal with this issue, one can consider clustering the profiles and calculating the mean separately for each cluster. To cluster the CP profiles, one may use standard methods like K-means or hierarchical clustering. The similarities between observations can be calculated based on the Euclidean distance between CP profiles.

Figure 17.2 illustrates an application of that approach to the random forest model `titanic_rf` (see Section 4.2.2) for 100 randomly selected instances (observations) from the Titanic dataset. The CP profiles for the *age* variable are marked in grey. It can be noted that they could be split into three clusters: one for a group of passengers with a substantial drop in the predicted survival probability for ages below 18 (with the average represented by the blue line), one with an almost linear decrease of the probability over the age (with the average represented by the red line), and one with almost constant predicted probability (with the average represented by the green line). The plot itself does not allow to identify the variables that may be linked with these clusters, but the additional exploratory analysis could be performed for this purpose.

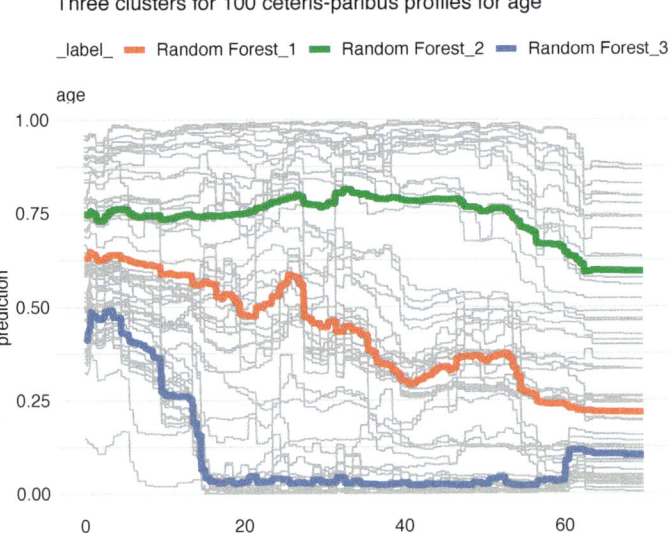

FIGURE 17.2 Clustered partial dependence profiles for *age* for the random forest model for 100 randomly selected observations from the Titanic dataset. Grey lines indicate ceteris-paribus profiles that are clustered into three groups with the average profiles indicated by the blue, green, and red lines.

17.3.3 Grouped partial-dependence profiles

It may happen that we can identify an explanatory variable that influences the shape of CP profiles for the explanatory variable of interest. The most obvious situation is when a model includes an interaction between the variable and another one. In that case, a natural approach is to investigate the PD profiles for the variable of interest within the groups of observations defined by the variable involved in the interaction.

Figure 17.3 illustrates an application of the approach to the random forest model `titanic_rf` (see Section 4.2.2) for 100 randomly selected instances (observations) from the Titanic dataset. The CP profiles for the explanatory-variable *age* are marked in grey. The red and blue lines present the PD profiles for females and males, respectively. The gender-specifc averages have different shapes: the predicted survival probability for females is more stable across different ages, as compared to males. Thus, the PD profiles clearly indicate an interaction between age and gender.

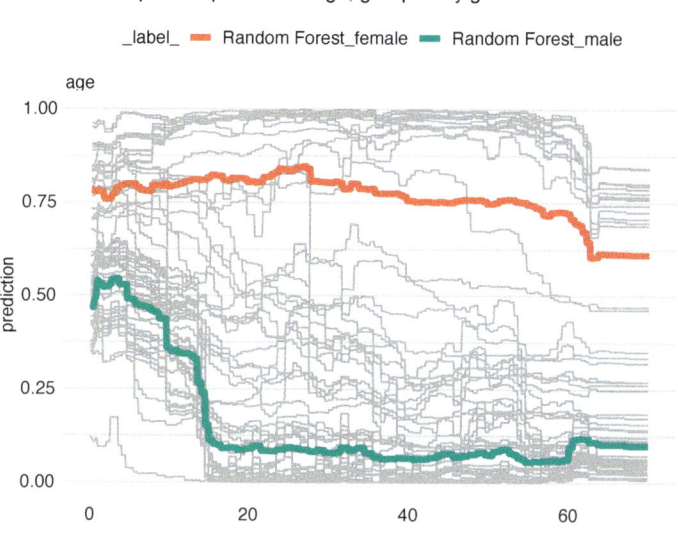

FIGURE 17.3 Partial-dependence profiles for two genders for the random forest model for 100 randomly selected observations from the Titanic dataset. Grey lines indicate ceteris-paribus profiles for *age*.

17.3.4 Contrastive partial-dependence profiles

Comparison of clustered or grouped PD profiles for a single model may provide important insight into, for instance, the stability of the model's predictions. PD profiles can also be compared between different models.

Figure 17.4 presents PD profiles for *age* for the random forest model (see Section 4.2.2) and the logistic regression model with splines for the Titanic data (see Section 4.2.2). The profiles are similar with respect to a general relationship between *age* and the predicted probability of survival (the younger the passenger, the higher chance of survival). However, the profile for the random forest model is flatter. The difference between both models is the largest at the left edge of the age scale. This pattern can be seen as expected because random forest models, in general, shrink predictions towards the average and they are not very good for extrapolation outside the range of values observed in the training dataset.

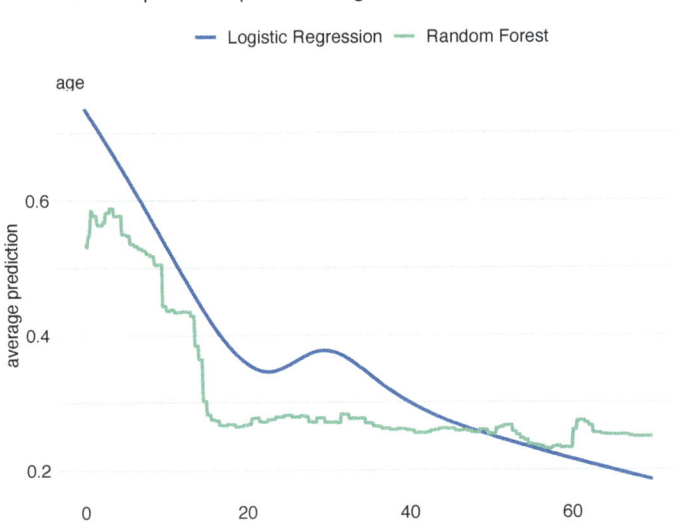

FIGURE 17.4 Partial-dependence profiles for *age* for the random forest (green line) and logistic regression (blue line) models for the Titanic dataset.

17.4 Example: apartment-prices data

In this section, we use PD profiles to evaluate performance of the random forest model (see Section 4.5.2) for the apartment-prices dataset (see Section 4.4). Recall that the goal is to predict the price per square meter of an apartment. In our illustration, we focus on two explanatory variables, *surface* and *construction year*. We consider the predictions for the training dataset `apartments`.

17.4.1 Partial-dependence profiles

Figure 17.5 presents CP profiles (grey lines) for 100 randomly-selected apartments together with the estimated PD profile (blue line) for *construction year* and *surface*.

PD profile for *surface* suggests an approximately linear relationship between the explanatory variable and the predicted price. On the other hand, PD profile for *construction year* is U-shaped: the predicted price is the highest for the very new and very old apartments. Note that, while the data were simulated, they were generated to reflect the effect of a lower quality of building materials used in rapid housing construction after the World War II.

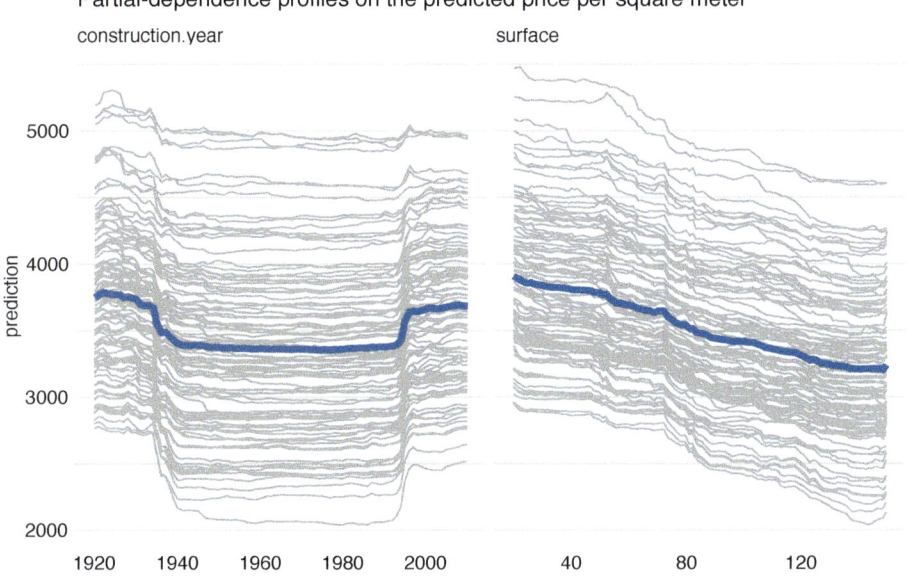

FIGURE 17.5 Ceteris-paribus and partial-dependence profiles for *construction year* and *surface* for 100 randomly-selected apartments for the random forest model for the apartment-prices dataset.

17.4.2 Clustered partial-dependence profiles

Almost all CP profiles for *construction year*, presented in Figure 17.5, seem to be U-shaped. The same shape is observed for the PD profile. One might want to confirm that the shape is, indeed, common for all the observations. The left-hand-side panel of Figure 17.6 presents clustered PD profiles for *construction year* for three clusters derived from the CP profiles presented in Figure 17.5. The three PD profiles differ slightly in the size of the oscillations at the edges, but they all are U-shaped. Thus, we could conclude that the overall PD profile adequately captures the shape of the CP profiles. Or, put differently, there is little evidence that there might be any strong interaction between year of construction and any other variable in the model. Similar conclusions can be drawn for the CP and PD profiles for *surface*, presented in the right-hand-side panel of Figure 17.6.

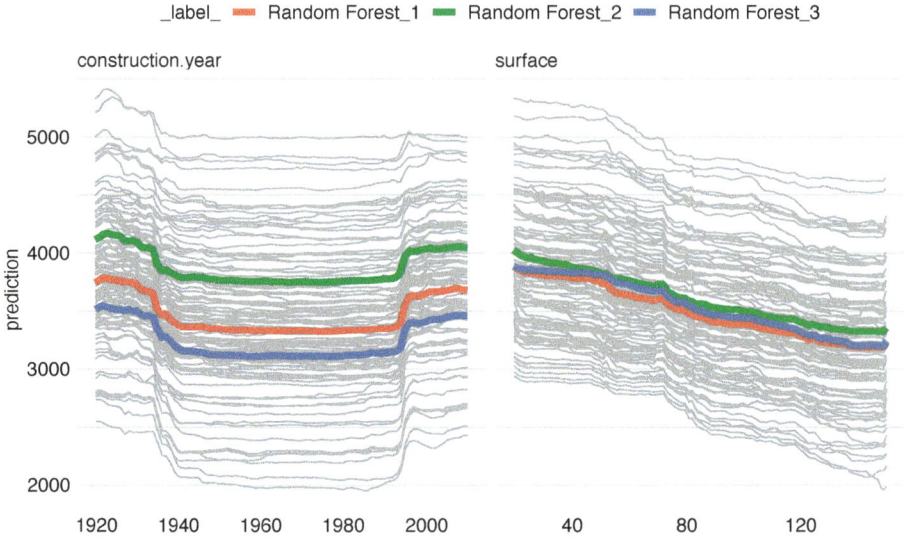

FIGURE 17.6 Ceteris-paribus (grey lines) and partial-dependence profiles (red, green, and blue lines) for three clusters for 100 randomly-selected apartments for the random forest model for the apartment-prices dataset. Left-hand-side panel: profiles for *construction year*. Right-hand-side panel: profiles for *surface*.

17.4.3   Grouped partial-dependence profiles

One of the categorical explanatory variables in the apartment prices dataset is *district*. We may want to investigate whether the relationship between the model's predictions and *construction year* and *surface* is similar for all districts. Toward this aim, we can use grouped PD profiles, for groups of apartments defined by districts.

Figure 17.7 shows PD profiles for *construction year* (left-hand-side panel) and *surface* (right-hand-side panel) for each district. Several observations are worth making. First, profiles for apartments in "Srodmiescie" (Downtown) are clearly much higher than for other districts. Second, the profiles are roughly parallel, indicating that the effects of *construction year* and *surface* are similar for each level of *district*. Third, the profiles appear to form three clusters, i.e., "Srodmiescie" (Downtown), three districts close to "Srodmiescie" (namely "Mokotow", "Ochota", and "Ursynow"), and the six remaining districts.

Partial-dependence profiles by district

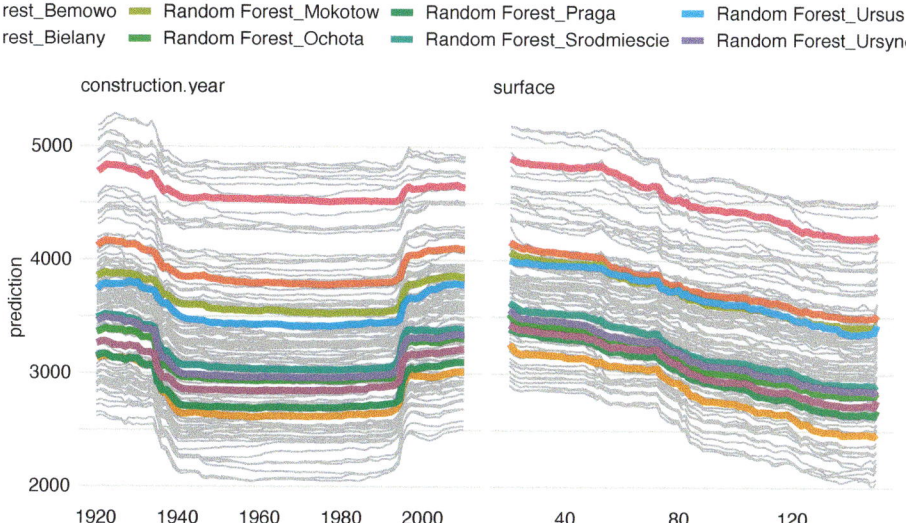

FIGURE 17.7 Partial-dependence profiles for separate districts for the random forest model for the apartment-prices dataset. Left-hand-side panel: profiles for *construction year*. Right-hand-side panel: profiles for *surface*.

17.4.4 Contrastive partial-dependence profiles

One of the main challenges in predictive modelling is to avoid overfitting. The issue is particularly important for flexible models, such as random forest models.

Figure 17.8 presents PD profiles for *construction year* (left-hand-side panel) and *surface* (right-hand-side panel) for the linear-regression model (see Section 4.5.1) and the random forest model. Several observations are worth making. The linear-regression model cannot, of course, accommodate the non-monotonic relationship between *construction year* and the price per square meter. However, for *surface*, both models support a linear relationship, though the slope of the line resulting from the linear regression is steeper. This may be seen as an expected difference, given that random forest models yield predictions that are shrunk towards the mean. Overall, we could cautiously conclude that there is not much evidence for overfitting of the more flexible random forest model.

Note that the non-monotonic relationship between *construction year* and the price per square meter might be the reason why the explanatory variable was found not to be important in the model in Section 16.6.

In Section 4.5.4, we mentioned that a proper model exploration may suggest a way to construct a model with improved performance, as compared to the random forest and linear-regression models. In this respect, it is worth observing that the profiles in Figure 17.8 suggest that both models miss some aspects of the data. In particular, the linear-regression model does not capture the U-shaped relationship between *construction year* and the price. On the other hand, the effect of *surface* on the apartment price seems to be underestimated by the random forest model. Hence, one could conclude that, by addressing the issues, one could improve either of the models, possibly with an improvement in predictive performance.

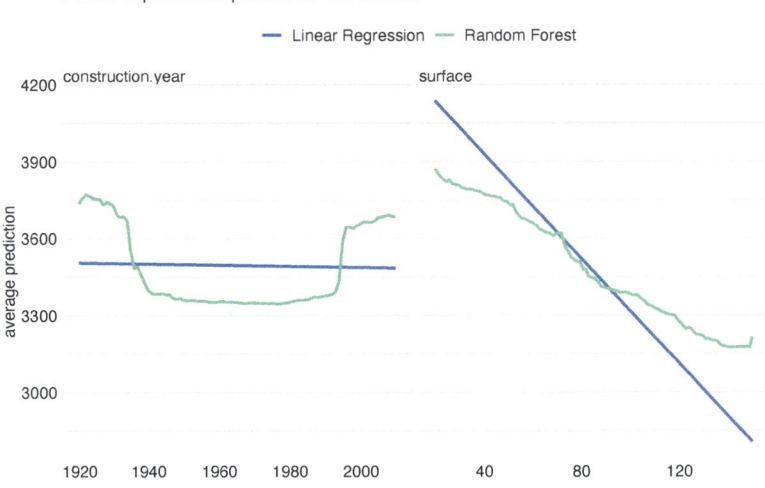

FIGURE 17.8 Partial-dependence profiles for the linear-regression and random forest models for the apartment-prices dataset. Left-hand-side panel: profiles for *construction year*. Right-hand-side panel: profiles for *surface*.

17.5   Pros and cons

PD profiles, presented in this chapter, offer a simple way to summarize the effect of a particular explanatory variable on the dependent variable. They are easy to explain and intuitive. They can be obtained for sub-groups of observations and compared across different models. For these reasons, they have gained in popularity and have been implemented in various software packages, including R and Python.

Given that the PD profiles are averages of CP profiles, they inherit the limitations of the latter. In particular, as CP profiles are problematic for correlated explanatory variables (see Section 10.5), PD profiles are also not suitable for that case, as they may offer a crude and potentially misleading summarization. An approach to deal with this issue will be discussed in the next chapter.

17.6 Code snippets for R

In this section, we present the `DALEX` package for R, which covers the methods presented in this chapter. It uses the `ingredients` package with various implementations of variable profiles. Similar functions can be found in packages `pdp` (Greenwell, 2017), `ALEPlots` (Apley, 2018), and `iml` (Molnar et al., 2018).

For illustration purposes, we use the random forest model `titanic_rf` (see Section 4.2.2) for the Titanic data. Recall that the model has been developed to predict the probability of survival from the sinking of the Titanic. We first retrieve the version of the `titanic` data with imputed missing values and the `titanic_rf` model-object via the `archivist` hooks, as listed in Section 4.2.7. Then we construct the explainer for the model by using function `explain()` from the `DALEX` package (see Section 4.2.6). Note that, beforehand, we have got to load the `randomForest` package, as the model was fitted by using function `randomForest()` from this package (see Section 4.2.2) and it is important to have the corresponding `predict()` function available.

```r
library("DALEX")
library("randomForest")
titanic_imputed <- archivist::aread("pbiecek/models/27e5c")
titanic_rf <- archivist::aread("pbiecek/models/4e0fc")
explainer_rf <- DALEX::explain(model = titanic_rf,
                               data = titanic_imputed[, -9],
                               y = titanic_imputed$survived,
                               label = "Random Forest")
```

17.6.1 Partial-dependence profiles

The function that allows computation of PD profiles in the `DALEX` package is `model_profile()`. The only required argument is `explainer`, which indicates the explainer-object (obtained with the help of the `explain()` function, see Section 4.2.6) for the model to be explained. The other useful arguments include:

- `variables`, a character vector providing the names of the explanatory variables, for which the profile is to be computed; by default, `variables = NULL`, in which case computations are performed for all numerical variables included in the model.
- `N`, the number of (randomly sampled) observations that are to be used for

the calculation of the PD profiles (N = 100 by default); N = NULL implies the use of the entire dataset included in the explainer-object.

- type, the type of the PD profile, with values "partial" (default), "conditional", and "accumulated".
- variable_type, a character string indicating whether calculations should be performed only for "numerical" (continuous) explanatory variables (default) or only for "categorical" variables.
- groups, the name of the explanatory variable that will be used to group profiles, with groups = NULL by default (in which case no grouping of profiles is applied).
- k, the number of clusters to be created with the help of the hclust() function, with k = NULL used by default and implying no clustering.

In the example below, we calculate the PD profile for *age* by applying the model_profile() function to the explainer-object for the random forest model. By default, the profile is based on 100 randomly selected observations.

```
pdp_rf <- model_profile(explainer = explainer_rf, variables = "age")
```

The resulting object of class model_profile contains the PD profile for *age*. By applying the plot() function to the object, we obtain a plot of the PD profile. Had we not used the variables argument, we would have obtained separate plots of PD profiles for all continuous explanatory variables.

```
library("ggplot2")
plot(pdp_rf) + ggtitle("Partial-dependence profile for age")
```

The resulting plot for *age* (not shown) corresponds to the one presented in Figure 17.4. It may slightly differ, as the two plots are based on different sets of (randomly selected) 100 observations from the Titanic dataset.

A PD profile can be plotted on top of CP profiles. This is a very useful feature if we want to check how well the former captures the latter. It is worth noting that, apart from the PD profile, the object created by the model_profile() function also contains the CP profiles for the selected observations and all explanatory variables included in the model. By specifying the argument geom = "profiles" in the plot() function, we add the CP profiles to the plot of the PD profile.

```
plot(pdp_rf, geom = "profiles") +
    ggtitle("Ceteris-paribus and partial-dependence profiles for age")
```

The resulting plot (not shown) is essentially the same as the one shown in the right-hand-side panel of Figure 17.1.

17.6.2   Clustered partial-dependence profiles

To calculate clustered PD profiles, we have got to cluster the CP profiles. Toward this aim, we use the k argument of the `model_profile()` function that specifies the number of clusters that are to be formed by the `hclust()` function. In the code below, we specify that three clusters are to be formed for profiles for *age*.

```
pdp_rf_clust <- model_profile(explainer = explainer_rf,
                              variables = "age", k = 3)
```

The clustered PD profiles can be plotted on top of the CP profiles by using the `geom = "profiles"` argument in the `plot()` function.

```
plot(pdp_rf_clust, geom = "profiles") +
    ggtitle("Clustered partial-dependence profiles for age")
```

The resulting plot (not shown) resembles the one shown for the random forest model in Figure 17.2. The only difference may stem from the fact that the two plots are based on a different set of (randomly selected) 100 observations from the Titanic dataset.

17.6.3   Grouped partial-dependence profiles

The `model_profile()` function admits the `groups` argument that allows constructing PD profiles for groups of observations defined by the levels of an explanatory variable. In the example below, we use the argument to obtain PD profiles for *age*, while grouping them by *gender*.

```
pdp_rf_gender <- model_profile(explainer = explainer_rf,
                               variables = "age", groups = "gender")
```

The grouped PD profiles can be plotted on top of the CP profiles by using the `geom = "profiles"` argument in the `plot()` function.

```
plot(pdp_rf_gender, geom = "profiles") +
    ggtitle("Partial-dependence profiles for age, grouped by gender")
```

The resulting plot (not shown) resembles the one shown in Figure 17.3.

17.6.4   Contrastive partial-dependence profiles

It may be of interest to compare PD profiles for several models. We will compare the random forest model with the linear-regression model `titanic_lmr` (see Section 4.2.1). For the latter, we first have got to load it via the **archivist** hook, as listed in Section 4.5.6. Then we construct the explainer for the model by using function `explain()`. Note that we first load the **rms** package, as the model was fitted by using function `lmr()` from this package (see Section 4.2.1) and it is important to have the corresponding `predict()` function available.

Finally, we apply the `model_profile()` function to compute CP profiles and the PD profile for *age* based on 100 randomly-selected observations from the Titanic dataset. We also repeat the calculations of the profiles for the random forest model.

```
library("rms")
titanic_lmr <- archivist::aread("pbiecek/models/58b24")
explainer_lmr <- DALEX::explain(model = titanic_lmr,
                                data = titanic_imputed[, -9],
                                y = titanic_imputed$survived,
                                label = "Logistic Regression")

pdp_lmr <- model_profile(explainer = explainer_lmr, variables = "age")
pdp_rf <- model_profile(explainer = explainer_rf, variables = "age")
```

To overlay the PD profiles for *age* for the two models in a single plot, we apply the `plot()` function to the `model_profile`-class objects for the two models that contain the profiles for *age*.

```
plot(pdp_rf, pdp_lmr) +
    ggtitle("Partial-dependence profiles for age for two models")
```

As a result, the profiles are plotted in a single plot. The resulting graph (not shown) is essentially the same as the one presented in Figure 17.4, with a possible difference due to the use of a different set of (randomly selected) 100 observations from the Titanic dataset.

17.7   Code snippets for Python

In this section, we use the `dalex` library for Python. The package covers all methods presented in this chapter. It is available on `pip` and `GitHub`. Similar functions can be found in library `PDPbox` (Jiangchun, 2018).

For illustration purposes, we use the `titanic_rf` random forest model for the Titanic data developed in Section 4.3.2. Recall that the model is developed to predict the probability of survival for passengers of Titanic.

In the first step, we create an explainer-object that will provide a uniform interface for the predictive model. We use the `Explainer()` constructor for this purpose.

```
import dalex as dx
titanic_rf_exp = dx.Explainer(titanic_rf, X, y,
                  label = "Titanic RF Pipeline")
```

The function that allows calculations of PD profiles is `model_profile()`. By default, it calculates profiles for all continuous variables. The other useful arguments include:

- `variables`, a `str`, `list`, `np.ndarray` or `pd.Series` providing the names of the explanatory variables, for which the profile is to be computed; by default computations are performed for all numerical variables included in the model.
- `N`, the number of (randomly sampled) observations that are to be used for the calculation of the PD profiles (`N = 300` by default); `N = None` implies the use of the entire dataset included in the explainer-object.
- `B`, the number of times (by default, 10) the entire procedure is to be repeated.
- `type`, the type of the PD profile, with values `'partial'` (default), `'conditional'`, and `'accumulated'`.
- `variable_type`, a character string indicating whether calculations should be performed only for `'numerical'` (continuous) explanatory variables (default) or only for `'categorical'` variables.
- `groups`, the name or list of names of the explanatory variable that will be used to group profiles, with `groups = None` by default (in which case no grouping of profiles is applied).

In the example below, we calculate the PD profiles for *age* and *fare* by applying the `model_profile()` function to the explainer-object for the random forest model. By default, the profile is based on 300 randomly selected observations.

```
pd_rf = titanic_rf_exp.model_profile(variables = ['age', 'fare'])
pd_rf.result
```

	vname	_label_	_x_	_yhat_	_ids_
0	age	Titanic RF Pipeline	0.166667	0.488955	0
1	age	Titanic RF Pipeline	0.905000	0.493257	0
2	age	Titanic RF Pipeline	1.643333	0.488888	0
3	age	Titanic RF Pipeline	2.381667	0.488888	0
4	age	Titanic RF Pipeline	3.120000	0.489874	0
...	...	...	...	...	...
197	fare	Titanic RF Pipeline	491.578272	0.446662	0
198	fare	Titanic RF Pipeline	496.698879	0.446662	0
199	fare	Titanic RF Pipeline	501.819486	0.446662	0
200	fare	Titanic RF Pipeline	506.940093	0.446662	0
201	fare	Titanic RF Pipeline	512.060700	0.446662	0

The results can be visualised by applying the `plot()` method. Figure 17.9 presents the created plot.

```
mp_rf.plot()
```

FIGURE 17.9 Partial-dependence profiles for *age* and *fare* for the random forest model for the Titanic data, obtained by using the `plot()` method in Python.

A PD profile can be plotted on top of CP profiles. This is a very useful feature if we want to check how well does the former capture the latter. By specifying the argument `geom = 'profiles'` in the `plot()` method, we add the CP profiles to the plot of the PD profile.

```
mp_rf.plot(geom = 'profiles')
```

The left-hand-side panel of the resulting plot (see Figure 17.10) is essentially the same as the one shown in the right-hand-side panel of Figure 17.1.

By default, the `model_profile()` function computes the PD profiles only for continuous explanatory variables. To obtain the profiles for categorical variables, in the code that follows we use the argument `variable_type='categorical'`. Additionally, in the call to the `plot()` method we indicate that we want to display the profiles only to variables *class* and *gender*.

```
mp_rf = titanic_rf_exp.model_profile( variable_type = 'categorical')
mp_rf.plot(variables = ['gender', 'class'])
```

The resulting plot is presented in Figure 17.11.

17.7.1 Grouped partial-dependence profiles

The `model_profile()` function admits the `groups` argument that allows constructing PD profiles for groups of observations defined by the levels of an explanatory variable. In the code below, we use the argument to compute the profiles for *age* and *fare*, while grouping them by *class*. Subsequently, we use

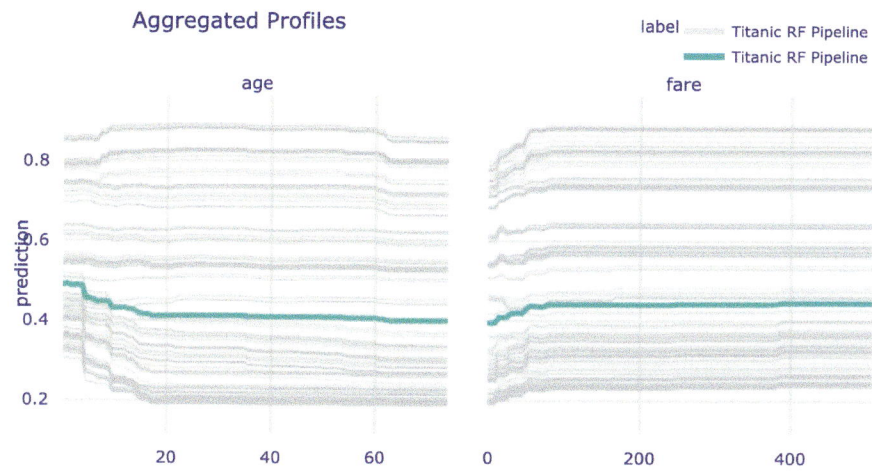

FIGURE 17.10 Partial-dependence profiles (blue) with corresponding ceteris-paribus profiles (grey) for *age* and *fare* for the random forest model for the Titanic data, obtained by using the `plot()` method in Python.

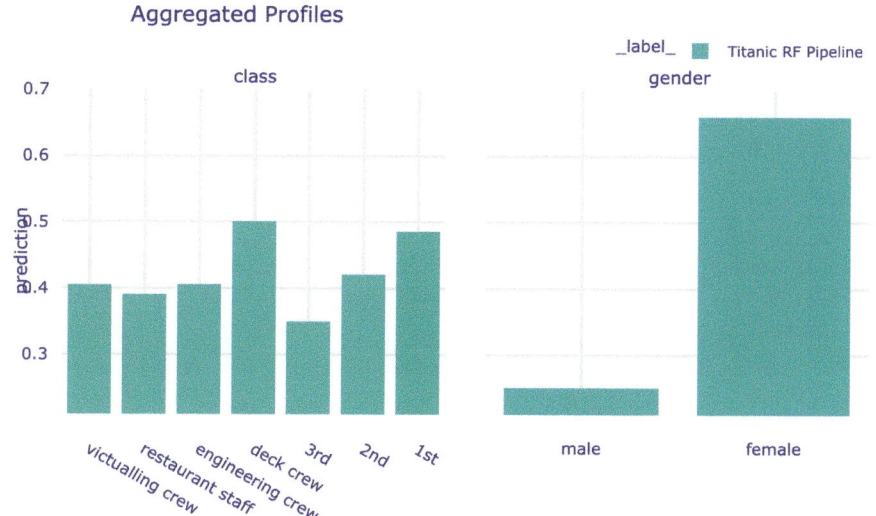

FIGURE 17.11 Partial-dependence profiles for *class* and *gender* for the random forest model for the Titanic data, obtained by using the `plot()` method in Python.

the `plot()` method to obtain a graphical presentation of the results. The resulting plot is presented in Figure 17.12.

```
mp_rf = titanic_rf_exp.model_profile(groups = 'class',
                                     variables = ['age', 'fare'])
mp_rf.plot()
```

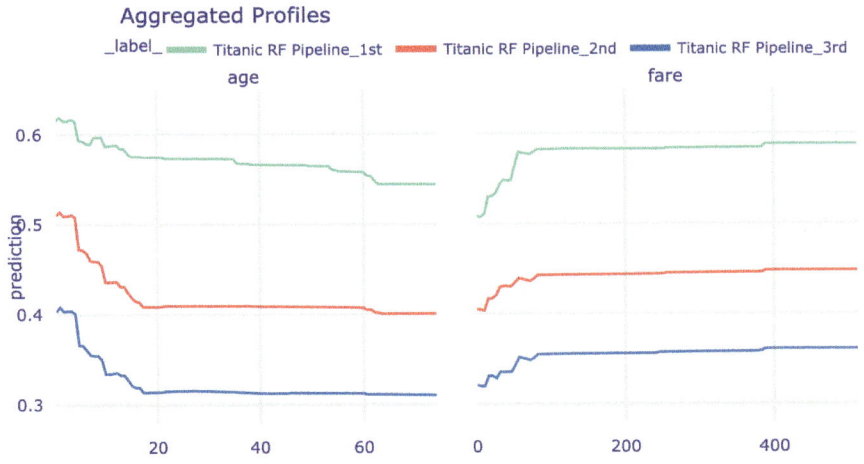

FIGURE 17.12 Partial-dependence profiles for *age* and *fare*, grouped by *class*, for the random forest model for the Titanic data, obtained by using the `plot()` method in Python.

17.7.2   Contrastive partial-dependence profiles

It may be of interest to compare PD profiles for several models. As an illustration, we will compare the random forest model with the logistic regression model `titanic_lr` (see Section 4.3.1). First, we have got to compute to the profiles for both models by using the `model_profile()` function.

```
pdp_rf = titanic_rf_exp.model_profile()
pdp_lr = titanic_lr_exp.model_profile()
```

Subsequently, we apply the `plot()` method to plot the profiles. Note that, in the code below, we use the `variables` argument to limit the display to variable *age* and *fare*.

```
pdp_rf.plot(pdp_lr, variables = ['age', 'fare'])
```

As a result, the profiles for *age* and *fare* are presented in a single plot. The resulting graph is presented in Figure 17.13).

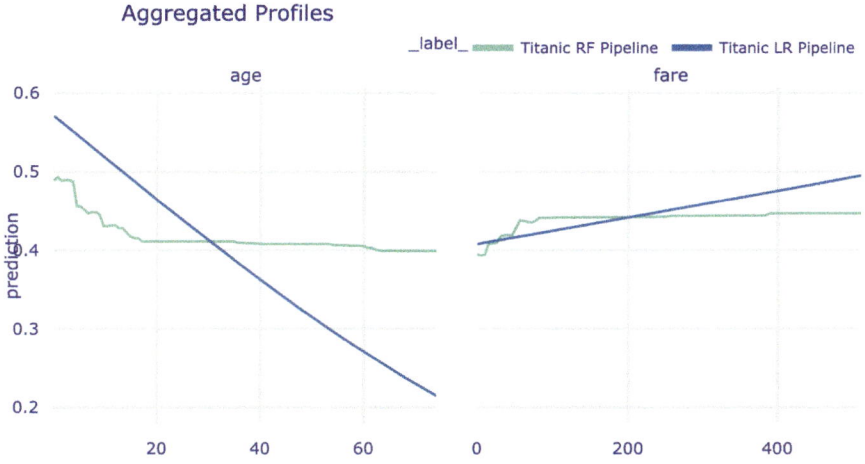

FIGURE 17.13 Partial-dependence profiles for *age* and *fare* for the random forest model and the logistic regression model for the Titanic data.

18

Local-dependence and Accumulated-local Profiles

18.1 Introduction

Partial-dependence (PD) profiles, introduced in the previous chapter, are easy to explain and interpret, especially given their estimation as the mean of ceteris-paribus (CP) profiles. However, as it was mentioned in Section 17.5, the profiles may be misleading if, for instance, explanatory variables are correlated. In many applications, this is the case. For example, in the apartment-prices dataset (see Section 4.4), one can expect that variables *surface* and *number of rooms* may be positively correlated, because apartments with a larger number of rooms usually also have a larger surface. Thus, in ceteris-paribus profiles, it is not realistic to consider, for instance, an apartment with five rooms and a surface of 20 square meters. Similarly, in the Titanic dataset, a positive association can be expected for the values of variables *fare* and *class*, as tickets in the higher classes are more expensive than in the lower classes.

In this chapter, we present accumulated-local profiles that address this issue. As they are related to local-dependence profiles, we introduce the latter first. Both approaches were proposed by Apley (2018).

18.2 Intuition

Let us consider the following simple linear model with two explanatory variables:

$$Y = X^1 + X^2 + \varepsilon = f(X^1, X^2) + \varepsilon, \tag{18.1}$$

where $\varepsilon \sim \mathcal{N}(0, 0.1^2)$.

For this model, the effect of X^1 for any value of X^2 is linear, i.e., it can be described by a straight line with the intercept equal to 0 and the slope equal to 1.

Assume that observations of explanatory variables X^1 ad X^2 are uniformly distributed over the unit square, as illustrated in the left-hand-side panel of Figure 18.1 for a set of 1000 observations. The right-hand-side panel of Figure 18.1 presents the scatter plot of the observed values of Y in function of X^1. The plot for X^2 is, essentially, the same and we do not show it.

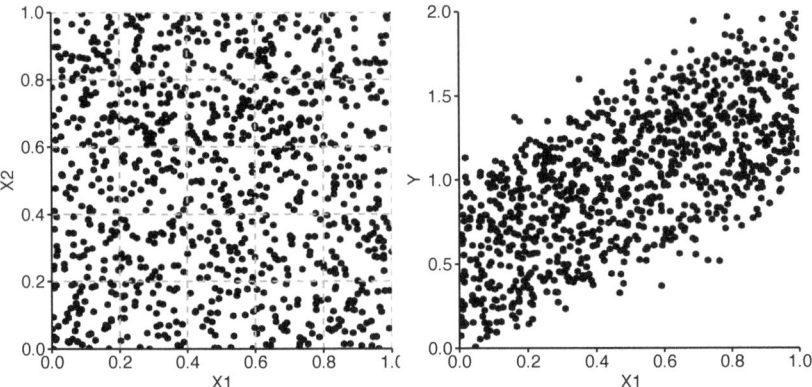

FIGURE 18.1 Observations of two explanatory variables uniformly distributed over the unit square (left-hand-side panel) and the scatter plot of the observed values of the dependent variable Y in function of X^1 (right-hand-side panel).

In view of the plot shown in the right-hand-side panel of Figure 18.1, we could consider using a simple linear model with X^1 and X^2 as explanatory variables. Assume, however, that we would like to analyze the data without postulating any particular parametric form of the effect of the variables. A naïve way would be to split the observed range of each of the two variables into, for instance, five intervals (as illustrated in the left-hand-side panel of Figure 18.1), and estimate the means of observed values of Y for the resulting 25 groups of observations. Table 18.1 presents the sample means (with rows and columns defined by the ranges of possible values of, respectively, X^1 and X^2).

TABLE 18.1: Sample means of Y for 25 groups of observations resulting from splitting the ranges of explanatory variables X^1 and X^2 into five intervals (see the left-hand-side panel of Figure 18.1).

	(0,0.2]	(0.2,0.4]	(0.4,0.6]	(0.6,0.8]	(0.8,1]
(0,0.2]	0.19	0.42	0.63	0.80	0.99
(0.2,0.4]	0.39	0.59	0.81	1.01	1.19
(0.4,0.6]	0.59	0.81	0.98	1.20	1.44

	(0,0.2]	(0.2,0.4]	(0.4,0.6]	(0.6,0.8]	(0.8,1]
(0.6,0.8]	0.76	1.00	1.20	1.40	1.58
(0.8,1]	1.01	1.22	1.38	1.58	1.77

Table 18.2 presents the number of observations for each of the sample means from Table 18.1.

TABLE 18.2: Number of observations for the sample means from Table 18.1.

	(0,0.2]	(0.2,0.4]	(0.4,0.6]	(0.6,0.8]	(0.8,1]	total
(0,0.2]	51	39	31	43	43	207
(0.2,0.4]	39	40	35	53	42	209
(0.4,0.6]	28	42	35	49	40	194
(0.6,0.8]	37	30	36	55	45	203
(0.8,1]	43	46	36	28	34	187
total	198	197	173	228	204	1000

By using this simple approach, we can compute the PD profile for X^1. Consider $X^1 = z$. To apply the estimator defined in (17.2), we need the predicted values $\hat{f}(z, x_i^2)$ for any observed value of $x_i^2 \in [0,1]$. As our observations are uncorrelated and fill-in the unit-square, we can use the suitable mean values for that purpose. In particular, for $z \in [0, 0.2]$, we get

$$\hat{g}_{PD}^1(z) = \frac{1}{1000} \sum_i \hat{f}(z, x_i^2) =$$
$$= (198 \times 0.19 + 197 \times 0.42 + 173 \times 0.63 +$$
$$228 \times 0.80 + 204 \times 1.00)/1000 = 0.6.$$

By following the same principle, for $z \in (0.2, 0.4]$, $(0.4, 0.6]$, $(0.6, 0.8]$, and $(0.8, 1]$ we get the values of 0.8, 1, 1.2, and 1.4, respectively. Thus, overall, we obtain a piecewise-constant profile with values that capture the (correct) linear effect of X^1 in model (18.1). In fact, by using, for instance, midpoints of the intervals for z, i.e., 0.1, 0.3, 0.5, 0.7, and 0.9, we could describe the profile by the linear function $0.5 + z$.

Assume now that we are given the data only from the regions on the diagonal of the unit square, as illustrated in the left-hand-side panel of Figure 18.2. In that case, the observed values of X^1 and X^2 are strongly correlated, with the estimated value of Pearson's correlation coefficient equal to 0.96. The

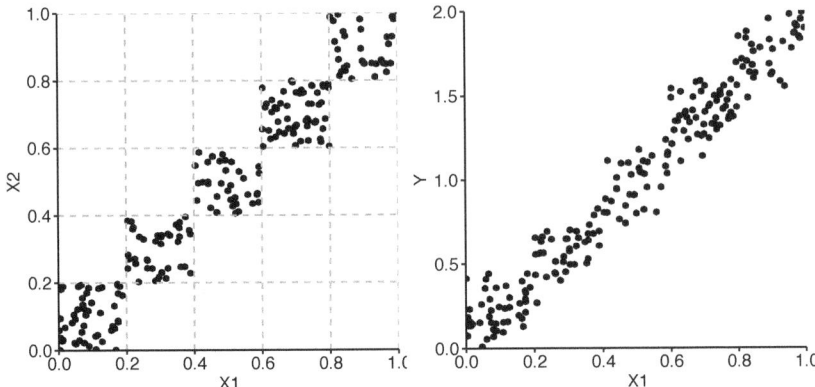

FIGURE 18.2 Correlated observations of two explanatory variables (left-hand-side panel) and the scatter plot of the observed values of the dependent variable Y in the function of X^1 (right-hand-side panel).

right-hand-side panel of Figure 18.2 presents the scatter plot of the observed values of Y in the function of X^1.

Now, the "naïve" modelling approach would amount to using only five sample means, as in the table below.

TABLE 18.3: Sample means of Y for five groups of observations (see the left-hand-side panel of Figure 18.2).

	(0,0.2]	(0.2,0.4]	(0.4,0.6]	(0.6,0.8]	(0.8,1]
(0,0.2]	0.19	NA	NA	NA	NA
(0.2,0.4]	NA	0.59	NA	NA	NA
(0.4,0.6]	NA	NA	0.98	NA	NA
(0.6,0.8]	NA	NA	NA	1.4	NA
(0.8,1]	NA	NA	NA	NA	1.77

When computing the PD profile for X^1, we now encounter the issue related to the fact that, for instance, for $z \in [0, 0.2]$, we have not got any observations and, hence, any sample mean for $x_i^2 > 0.2$. To overcome this issue, we could extrapolate the predictions (i.e., mean values) obtained for other intervals of z. That is, we could assume that, for $x_i^2 \in (0.2, 0.4]$, the prediction is equal to 0.59, for $x_i^2 \in (0.4, 0.6]$ it is equal to 0.98, and so on. This leads to the following value of the PD profile for $z \in [0, 0.2]$:

$$\hat{g}_{PD}^1(z) = \frac{1}{207} \sum_i \hat{f}(z, x_i^2) =$$

$$= \frac{1}{1000}(198 \times 0.19 + 197 \times 0.59 + 173 \times 0.99 +$$
$$228 \times 1.40 + 204 \times 1.77) = 1. \tag{18.2}$$

This is a larger value than 0.6 computed in (18.2) for the uncorrelated data. The reason is the extrapolation: for instance, for $z \in [0, 0.2]$ and $x_i^2 \in (0.6, 0.8]$, we use 1.40 as the predicted value of Y. However, Table 18.1 indicates that the sample mean for those observations is equal to 0.80.

In fact, by using the same extrapolation principle, we get $\hat{g}_{PD}^1(z) = 1$ also for $z \in (0.2, 0.4]$, $(0.4, 0.6]$, $(0.6, 0.8]$, and $(0.8, 1]$. Thus, the obtained profile indicates no effect of X^1, which is clearly a wrong conclusion.

While the modelling approach presented in the example above may seem to be simplistic, it does illustrate the issue that would also appear for other flexible modelling methods like, for instance, regression trees. In particular, the left-hand-side panel of Figure 18.3 presents a regression tree fitted to the data shown in Figure 18.2 by using function tree() from the R package tree. The right-hand-side panel of Figure 18.3 presents the corresponding split of the observations. According to the model, the predicted value of Y for the observations in the region $x^1 \in [0, 0.2]$ and $x^2 \in [0.8, 1]$ would be equal to 1.74. This extrapolation implies a substantial overestimation, as the true expected value of Y in the region is equal to 1. Note that the latter is well estimated by the sample mean equal to 0.99 (see Table 18.1) in the case of the uncorrelated data shown in Figure 18.1.

The PD profile for X^1 for the regression tree would be equal to 0.2, 0.8, and 1.5 for $z \in [0, 0.2]$, $(0.2, 0.6]$, and $(0.6, 1]$, respectively. It does show an effect of X^1, but if we used midpoints of the intervals for z, i.e., 0.1, 0.4, and 0.8, we could (approximately) describe the profile by the linear function $2z$, i.e., with a slope larger than (the true value of) 1.

The issue stems from the fact that, in the definition (17.1) of the PD profile, the expected value of model predictions is computed by using the marginal distribution of X^2, which disregards the value of X^1. Clearly, this is an issue when the explanatory variables are correlated. This observation suggests a modification: instead of the marginal distribution, one might use the conditional distribution of X^2 given X^1, because it reflects the association between the two variables. The modification leads to the definition of an LD profile.

It turns out, however, that the modification does not fully address the issue of correlated explanatory variables. As argued by Apley and Zhu (2020), if an explanatory variable is correlated with some other variables, the LD profile

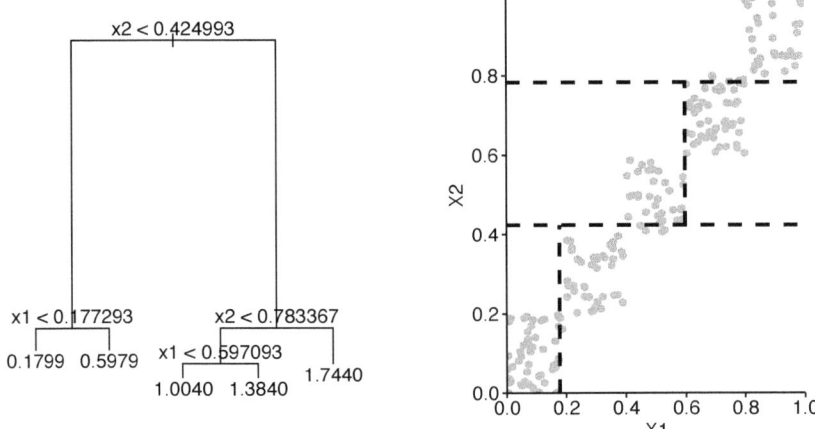

FIGURE 18.3 Results of fitting of a regression tree to the data shown in Figure 18.2 (left-hand-side panel) and the corresponding split of the observations of the two explanatory variables (right-hand-side panel).

for the variable will still capture the effect of the other variables. This is because the profile is obtained by marginalizing over (in fact, ignoring) the remaining variables in the model, which results in an effect similar to the "omitted variable" bias in linear regression. Thus, in this respect, LD profiles share the same limitation as PD profiles. To address the limitation, Apley and Zhu (2020) proposed the concept of local-dependence effects and accumulated-local (AL) profiles.

18.3   Method

18.3.1   Local-dependence profile

Local-dependence (LD) profile for model $f()$ and variable X^j is defined as follows:

$$g_{LD}^{f,j}(z) = E_{\underline{X}^{-j}|X^j=z}\left\{f\left(\underline{X}^{j|=z}\right)\right\}. \tag{18.3}$$

Thus, it is the expected value of the model predictions over the conditional distribution of $\underline{X}^{-j}$ given $X^j = z$, i.e., over the joint distribution of all explanatory variables other than X^j conditional on the value of the latter variable set to z. Or, in other words, it is the expected value of the CP profiles for X^j, defined in (10.1), over the conditional distribution of $\underline{X}^{-j}|X^j = z$.

As proposed by Apley and Zhu (2020), LD profile can be estimated as follows:

$$\hat{g}_{LD}^j(z) = \frac{1}{|N_j|} \sum_{k \in N_j} f\left(\underline{x}_k^{j|=z}\right), \qquad (18.4)$$

where N_j is the set of observations with the value of X^j "close" to z that is used to estimate the conditional distribution of $\underline{X}^{-j}|X^j = z$.

Note that, in general, the estimator given in (18.4) is neither smooth nor continuous at boundaries between subsets N_j. A smooth estimator for $g_{LD}^{f,j}(z)$ can be defined as follows:

$$\tilde{g}_{LD}^j(z) = \frac{1}{\sum_k w_k(z)} \sum_{i=1}^{n} w_i(z) f\left(\underline{x}_i^{j|=z}\right), \qquad (18.5)$$

where weights $w_i(z)$ capture the distance between z and x_i^j. In particular, for a categorical variable, we may just use the indicator function $w_i(z) = 1_{z=x_i^j}$, while for a continuous variable we may use the Gaussian kernel:

$$w_i(z) = \phi(z - x_i^j, 0, s), \qquad (18.6)$$

where $\phi(y, 0, s)$ is the density of a normal distribution with mean 0 and standard deviation s. Note that s plays the role of a smoothing factor.

As already mentioned in Section 18.2, if an explanatory variable is correlated with some other variables, the LD profile for the variable will capture the effect of all of the variables. For instance, consider model (18.1). Assume that X^1 has a uniform distribution on $[0, 1]$ and that $X^1 = X^2$, i.e., explanatory variables are perfectly correlated. In that case, the LD profile for X^1 is given by

$$g_{LD}^1(z) = E_{X^2|X^1=z}(z + X^2) = z + E_{X^2|X^1=z}(X^2) = 2z.$$

Hence, it suggests an effect of X^1 twice larger than the correct one.

To address the limitation, AL profiles can be used. We present them in the next section.

18.3.2   Accumulated-local profile

Consider model $f()$ and define

$$q^j(\underline{u}) = \left\{ \frac{\partial f(\underline{x})}{\partial x^j} \right\}_{\underline{x}=\underline{u}}.$$

Accumulated-local (AL) profile for model $f()$ and variable X^j is defined as follows:

$$g_{AL}^j(z) = \int_{z_0}^{z} \left[E_{\underline{X}^{-j}|X^j=v} \left\{ q^j(\underline{X}^{j|=v}) \right\} \right] dv + c, \qquad (18.7)$$

where z_0 is a value close to the lower bound of the effective support of the distribution of X^j and c is a constant, usually selected so that $E_{X^j}\left\{ g_{AL}^j(X^j) \right\} = 0$.

To interpret (18.7), note that $q^j(\underline{x}^{j|=v})$ describes the local effect (change) of the model due to X^j. Or, to put it in other words, $q^j(\underline{x}^{j|=v})$ describes how much the CP profile for X^j changes at $(x^1, \ldots, x^{j-1}, v, x^{j+1}, \ldots, x^p)$. This effect (change) is averaged over the "relevant" (according to the conditional distribution of $\underline{X}^{-j}|X^j$) values of $\underline{x}^{-j}$ and, subsequently, accumulated (integrated) over values of v up to z. As argued by Apley and Zhu (2020), the averaging of the local effects allows avoiding the issue, present in the PD and LD profiles, of capturing the effect of other variables in the profile for a particular variable in additive models (without interactions). To see this, one can consider the approximation

$$f(\underline{x}^{j|=v+dv}) - f(\underline{x}^{j|=v}) \approx q^j(\underline{x}^{j|=v})dv,$$

and note that the difference $f(\underline{x}^{j|=v+dv}) - f(\underline{v}^{j|=v})$, for a model without interaction, effectively removes the effect of all variables other than X^j.

For example, consider model (18.1). In that case, $f(x^1, x^2) = x^1 + x^2$ and $q^1(\underline{u}) = 1$. Thus,

$$f(u + du, x_2) - f(u, x_2) = (u + du + x^2) - (u + x^2) = du = q^1(u)du.$$

Consequently, irrespective of the joint distribution of X^1 and X^2 and upon setting $c = z_0$, we get

$$g_{AL}^1(z) = \int_{z_0}^{z} \left\{ E_{X^2|X^1=v}(1) \right\} dv + z_0 = z.$$

To estimate an AL profile, one replaces the integral in (18.7) by a summation and the derivative with a finite difference (Apley and Zhu, 2020). In particular, consider a partition of the range of observed values x_i^j of variable X^j into K intervals $N_j(k) = \left(z_{k-1}^j, z_k^j\right]$ $(k = 1, \ldots, K)$. Note that z_0^j can be chosen just below $\min(x_1^j, \ldots, x_N^j)$ and $z_K^j = \max(x_1^j, \ldots, x_N^j)$. Let $n_j(k)$ denote the number of observations x_i^j falling into $N_j(k)$, with $\sum_{k=1}^{K} n_j(k) = n$. An estimator of the AL profile for variable X^j can then be constructed as follows:

$$\widehat{g}_{AL}^j(z) = \sum_{k=1}^{k_j(z)} \frac{1}{n_j(k)} \sum_{i:x_i^j \in N_j(k)} \left\{ f\left(\underline{x}_i^{j|=z_k^j}\right) - f\left(\underline{x}_i^{j|=z_{k-1}^j}\right) \right\} - \hat{c}, \quad (18.8)$$

where $k_j(z)$ is the index of interval $N_j(k)$ in which z falls, i.e., $z \in N_j\{k_j(z)\}$, and $\hat{c}$ is selected so that $\sum_{i=1}^{n} \widehat{g}_{AL}^{f,j}(x_i^j) = 0$.

To interpret (18.8), note that difference $f\left(\underline{x}_i^{j|=z_k^j}\right) - f\left(\underline{x}_i^{j|=z_{k-1}^j}\right)$ corresponds to the difference of the CP profile for the i-th observation at the limits of interval $N_j(k)$. These differences are then averaged across all observations for which the observed value of X^j falls into the interval and are then accumulated.

Note that, in general, $\widehat{g}_{AL}^{f,j}(z)$ is not smooth at the boundaries of intervals $N_j(k)$. A smooth estimate can obtained as follows:

$$\widetilde{g}_{AL}^j(z) = \sum_{k=1}^{K} \left[\frac{1}{\sum_l w_l(z_k)} \sum_{i=1}^{N} w_i(z_k) \left\{ f\left(\underline{x}_i^{j|=z_k}\right) - f\left(\underline{x}_i^{j|=z_k - \Delta}\right) \right\} \right] - \hat{c},$$
$$(18.9)$$

where points z_k $(k = 0, \ldots, K)$ form a uniform grid covering the interval (z_0, z) with step $\Delta = (z - z_0)/K$, and weight $w_i(z_k)$ captures the distance between point z_k and observation x_i^j. In particular, we may use similar weights as in case of (18.5).

18.3.3 Dependence profiles for a model with interaction and correlated explanatory variables: an example

In this section, we illustrate in more detail the behavior of PD, LD, and AL profiles for a model with an interaction between correlated explanatory variables. In particular, let us consider the following simple model for two explanatory variables:

$$f(X^1, X^2) = (X^1 + 1) \cdot X^2. \quad (18.10)$$

Moreover, assume that explanatory variables X^1 and X^2 are uniformly distributed over the interval $[-1, 1]$ and perfectly correlated, i.e., $X^2 = X^1$. Suppose that we have got a dataset with eight observations as in Table 18.4. Note that, for both X^1 and X^2, the sum of all observed values is equal to 0.

TABLE 18.4: A sample of eight observations.

i	1	2	3	4	5	6	7	8
X^1	-1	-0.71	-0.43	-0.14	0.14	0.43	0.71	1
X^2	-1	-0.71	-0.43	-0.14	0.14	0.43	0.71	1
y	0	-0.2059	-0.2451	-0.1204	0.1596	0.6149	1.2141	2

Note that PD, LD, AL profiles describe the effect of a variable in isolation from the values of other variables. In model (18.10), the effect of variable X^1 depends on the value of variable X^2. For models with interactions, it is subjective to define what would be the "true" main effect of variable X^1. Complex predictive models often have interactions. By examining the case of model (18.10), we will provide some intuition on how PD, LD and AL profiles may behave in such cases.

Let us explicitly express the CP profile for X^1 for model (18.10):

$$h_{CP}^1(z) = f(z, X^2) = (z + 1) \cdot X^2. \tag{18.11}$$

By allowing z to take any value in the interval $[-1, 1]$, we get the CP profiles as straight lines with the slope equal to the value of variable X^2. Hence, for instance, the CP profile for observation $(-1, -1)$ is a straight line with the slope equal to -1. The CP profiles for the eight observations, from Table 18.4 are presented in panel A of Figure 18.4.

Recall that the PD profile for X^j, defined in equation (17.1), is the expected value, over the joint distribution of all explanatory variables other than X^j, of the model predictions when X^j is set to z. This leads to the estimation of the profile by taking the average of CP profiles for X^j, as given in (17.2).

In our case, this implies that the PD profile for X^1 is the expected value of the model predictions over the distribution of X^2, i.e., over the uniform distribution on the interval $[-1, 1]$. Thus, the PD profile is estimated by taking the average of the CP profiles, given by (18.11), at each value of z in $[-1, 1]$:

$$\hat{g}_{PD}^1(z) = \frac{1}{8} \sum_{i=1}^{8} (z + 1) \cdot X_i^2 = \frac{z + 1}{8} \sum_{i=1}^{8} X_i^2 = 0. \tag{18.12}$$

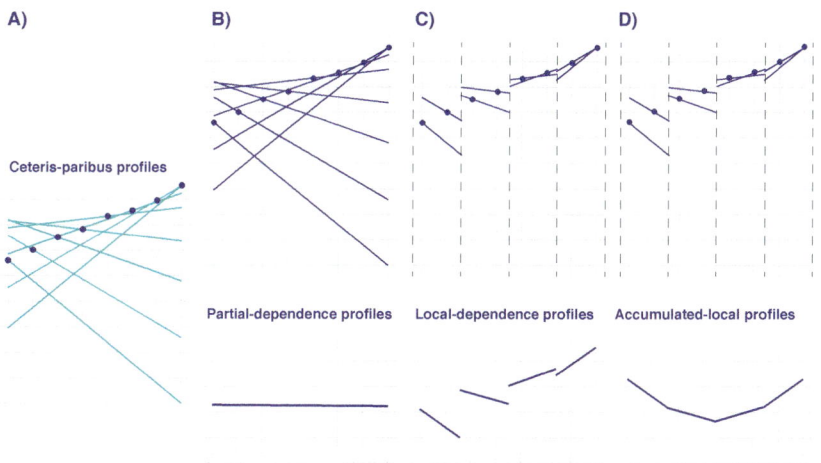

FIGURE 18.4 Partial-dependence (PD), local-dependence (LD), and accumulated-local (AL) profiles for model (18.10). Panel A: ceteris-paribus (CP) profiles for eight observations from Table 18.4. Panel B: entire CP profiles (top) contribute to calculation of the corresponding PD profile (bottom). Panel C: only parts of the CP profiles (top), close to observations of interest, contribute to the calculation of the corresponding LD profile (bottom). Panel D: only parts of the CP profiles (top) contribute to the calculation of the corresponding AL profile (bottom).

As a result, the PD profile for X^1 is estimated as a horizontal line at 0, as seen in the bottom part of Panel B of Figure 18.4.

Since the X^1 and X^2 variables are correlated, it can be argued that we should not include entire CP profiles in the calculation of the PD profile, but only parts of them. In fact, for perfectly correlated explanatory variables, the CP profile for the i-th observation should actually be undefined for any values of z different from x_i^2.

The estimated horizontal PD profile results from using the marginal distribution of X^2, which disregards the value of X^1, in the definition of the profile. This observation suggests a modification: instead of the marginal distribution, one might consider the conditional distribution of X^2 given X^1. The modification leads to the definition of LD profile.

For the data from Table 18.4, the conditional distribution of X^2, given $X^1 = z$, is just a probability mass of 1 at z. Consequently, for model (18.10), the LD profile for X^1 and any $z \in [-1, 1]$ is given by

$$g_{LD}^1(z) = z \cdot (z + 1). \tag{18.13}$$

The bottom part of panel C of Figure 18.4 presents the LD profile estimated by applying estimator (18.3), in which the conditional distribution was calculated by using four bins with two observations each (shown in the top part of the panel). The LD profile shows the average of predictions over the conditional distribution. Part of the average can be attributed to the effect of the correlated variable X^2. AL profile shows the net effect of X^1 variable.

By using definition (18.7), the AL profile for model (18.10) is given by

$$
\begin{aligned}
g^1_{AL}(z) &= \int_{-1}^{z} E\left[\frac{\partial f(X^1, X^2)}{\partial X^1}\bigg| X^1 = v\right] dv \\
&= \int_{-1}^{z} E\left[X^2 | X^1 = v\right] dv = \int_{-1}^{z} v\, dv = (z^2 - 1)/2. \tag{18.14}
\end{aligned}
$$

The bottom part of panel D of Figure 18.4 presents the AL profile estimated by applying estimator (18.7), in which the range of observed values of X^1 was split into four intervals with two observations each.

It is clear that PD, LD and AL profiles show different aspects of the model. In the analyzed example of model (18.10), we obtain three different explanations of the effect of variable X^1.

In practice, explanatory variables are typically correlated and complex predictive models are usually not additive. Therefore, when analyzing any model, it is worth checking how much do the PD, LD, and AL profiles differ. And if so, look for potential causes. Correlations can be detected at the stage of data exploration. Interactions can be noted by looking at individual CP profiles.

18.4   Example: apartment-prices data

In this section, we use PD, LD, and AL profiles to evaluate performance of the random forest model `apartments_rf` (see Section 4.5.2) for the apartment-prices dataset (see Section 4.4). Recall that the goal is to predict the price per square meter of an apartment. In our illustration, we focus on two explanatory variables, *surface* and *number of rooms*, as they are correlated (see Figure 4.9).

Figure 18.5 shows the three types of profiles for both variables estimated according to formulas (17.2), (18.5), and (18.9). As we can see from the plots, the profiles calculated with different methods are different. The LD profiles are steeper than the PD profiles. This is because, for instance, the effect of *surface* includes the effect of other correlated variables, including *number of rooms*. The AL profile eliminates the effect of correlated variables. Since the

AL and PD profiles are parallel to each other, they suggest that the model is additive for these two explanatory variables.

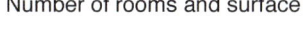

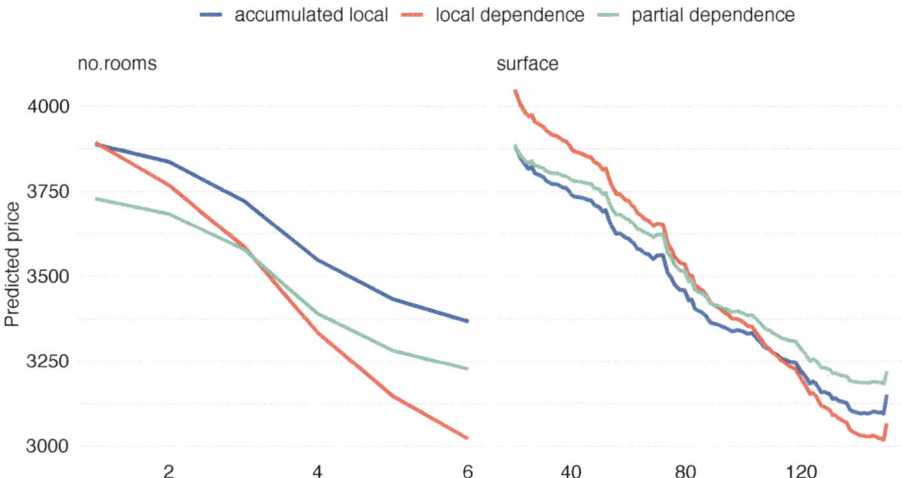

FIGURE 18.5 Partial-dependence, local-dependence, and accumulated-local profiles for the random forest model for the apartment-prices dataset.

18.5   Pros and cons

The LD and AL profiles, described in this chapter, are useful to summarize the influence of an explanatory variable on a model's predictions. The profiles are constructed by using the CP profiles introduced in Chapter 10, but they differ in how the CP profiles for individual observations are summarized.

When explanatory variables are independent and there are no interactions in the model, the CP profiles are parallel and their mean, i.e., the PD profile introduced in Chapter 17, adequately summarizes them.

When the model is additive, but an explanatory variable is correlated with some other variables, neither PD nor LD profiles will properly capture the effect of the explanatory variable on the model's predictions. However, the AL profile will provide a correct summary of the effect.

When there are interactions in the model, none of the profiles will provide a correct assessment of the effect of any explanatory variable involved in the interaction(s). This is because the profiles for the variable will also include the effect of other variables. Comparison of PD, LD, and AL profiles may help in identifying whether there are any interactions in the model and/or whether explanatory variables are correlated. When there are interactions, they may be explored by using a generalization of the PD profiles for two or more dependent variables (Apley and Zhu, 2020).

18.6   Code snippets for R

In this section, we present the DALEX package for R, which covers the methods presented in this chapter. In particular, it includes wrappers for functions from the ingredients package (Biecek et al., 2019). Note that similar functionalities can be found in package ALEPlots (Apley, 2018) or iml (Molnar et al., 2018).

For illustration purposes, we use the random forest model apartments_rf (see Section 4.5.2) for the apartment prices dataset (see Section 4.4). Recall that the goal is to predict the price per square meter of an apartment. In our illustration, we focus on two explanatory variables, *surface* and *number of rooms*.

We first load the model-object via the archivist hook, as listed in Section 4.5.6. Then we construct the explainer for the model by using the function explain() from the DALEX package (see Section 4.2.6). Note that, beforehand, we have got to load the randomForest package, as the model was fitted by using function randomForest() from this package (see Section 4.2.2) and it is important to have the corresponding predict() function available.

```
library("DALEX")
library("randomForest")
apartments_rf <- archivist::aread("pbiecek/models/fe7a5")
explainer_apart_rf <- DALEX::explain(model = apartments_rf,
                         data    = apartments_test[,-1],
                         y       = apartments_test$m2.price,
                         label   = "Random Forest")
```

The function that allows the computation of LD and AL profiles in the DALEX package is model_profile(). Its use and arguments were described in Section 17.6. LD profiles are calculated by specifying argument type = "conditional". In the example below, we also use the variables argument to calculate the profile only for the explanatory variables *surface* and *no.rooms*. By default, the profile is based on 100 randomly selected observations.

```
ld_rf <- model_profile(explainer = explainer_apart_rf,
                   type      = "conditional",
                   variables = c("no.rooms", "surface"))
```

The resulting object of class "model_profile" contains the LD profiles for both explanatory variables. By applying the `plot()` function to the object, we obtain separate plots of the profiles.

```
plot(ld_rf) +
  ggtitle("Local-dependence profiles for no. of rooms and surface", "")
```

The resulting plot is shown in Figure 18.6. The profiles essentially correspond to those included in Figure 18.5.

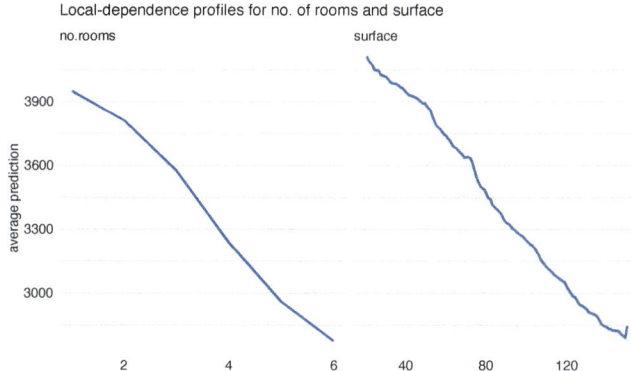

FIGURE 18.6 Local-dependence profiles for the random forest model and explanatory variables *no.rooms* and *surface* for the apartment-prices dataset.

AL profiles are calculated by applying function `model_profile()` with the additional argument `type = "accumulated"`. In the example below, we also use the `variables` argument to calculate the profile only for the explanatory variables *surface* and *no.rooms*.

```
al_rf <- model_profile(explainer = explainer_apart_rf,
                         type = "accumulated",
                       variables = c("no.rooms", "surface"))
```

By applying the `plot()` function to the object, we obtain separate plots of the AL profiles for *no.rooms* and *surface*. They are presented in Figure 18.7. The profiles essentially correspond to those included in Figure 18.5.

```
plot(al_rf) +
  ggtitle("Accumulated-local profiles for no. of rooms and surface", "")
```

Function `plot()` allows including all plots in a single graph. We will show how to apply it in order to obtain Figure 18.5. Toward this end, we have got to create PD profiles first (see Section 17.6). We also modify the labels of the PD, LD, and AL profiles contained in the `agr_profiles` components of the "model_profile"-class objects created for the different profiles.

```
pd_rf <- model_profile(explainer = explainer_apart_rf,
```

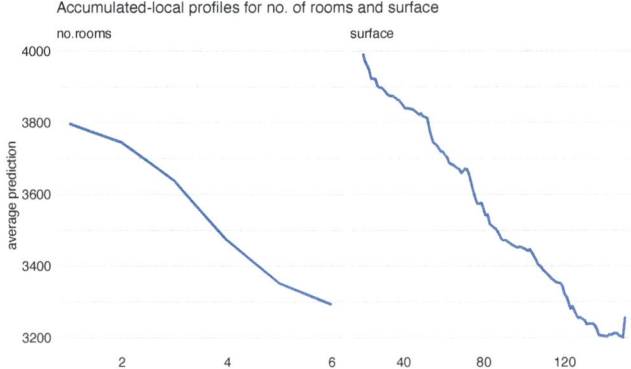

FIGURE 18.7 Accumulated-local profiles for the random forest model and explanatory variables *no.rooms* and *surface* for the apartment-prices dataset.

```
                  type = "partial",
        variables = c("no.rooms", "surface"))
```

Subsequently, we simply apply the `plot()` function to the `agr_profiles` components of the "model_profile"-class objects for the different profiles (see Section 17.6).

```
plot(pd_rf, ld_rf, al_rf)
```

The resulting plot (not shown) is essentially the same as the one presented in Figure 18.5, with a possible difference due to the use of a different set of (randomly selected) 100 observations from the apartment-prices dataset.

18.7 Code snippets for Python

In this section, we use the `dalex` library for Python. The package covers all methods presented in this chapter. It is available on `pip` and `GitHub`.

For illustration purposes, we use the `titanic_rf` random forest model for the Titanic data developed in Section 4.3.2. Recall that the model is developed to predict the probability of survival for passengers of Titanic.

In the first step, we create an explainer-object that will provide a uniform interface for the predictive model. We use the `Explainer()` constructor for this purpose.

```
import dalex as dx
titanic_rf_exp = dx.Explainer(titanic_rf, X, y,
                    label = "Titanic RF Pipeline")
```

The function that allows calculations of LD profiles is `model_profile()`. It was already introduced in Section 17.7. By defaut, it calculates PD profiles. To obtain LD profiles, the `type = 'conditional'` should be used.

In the example below, we calculate the LD profile for *age* and *fare* by applying the `model_profile()` function to the explainer-object for the random forest model while specifying `type = 'conditional'`. Results are stored in the `ld_rf.result` field.

```
ld_rf = titanic_rf_exp.model_profile(type = 'conditional')
ld_rf.result['_label_'] = 'LD profiles'
ld_rf.result
```

	vname	_label_	_grid_	_x_	_yhat_	_ids_
0	age	LD profiles	0	0.166667	0.468324	0
1	age	LD profiles	1	0.905000	0.472722	0
2	age	LD profiles	2	1.643333	0.468534	0
3	age	LD profiles	3	2.381667	0.468628	0
4	age	LD profiles	4	3.120000	0.469591	0
...	...	...	...	...	...	...
399	sibsp	LD profiles	96	7.680000	0.385616	0
400	sibsp	LD profiles	97	7.760000	0.384816	0
401	sibsp	LD profiles	98	7.840000	0.384031	0
402	sibsp	LD profiles	99	7.920000	0.383262	0
403	sibsp	LD profiles	100	8.000000	0.382509	0

Results can be visualised by using the `plot()` method. Note that, in the code below, we use the `variables` argument to display the LD profiles only for *age* and *fare*. The resulting plot is presented in Figure 18.8.

```
ld_rf.plot(variables = ['age', 'fare'])
```

In order to calculate the AL profiles for *age* and *fare*, we apply the `model_profile()` function with the `type = 'accumulated'` option.

```
al_rf = titanic_rf_exp.model_profile(type = 'accumulated')
al_rf.result['_label_'] = 'AL profiles'
```

FIGURE 18.8 Local-dependence profiles for *age* and *fare* for the random forest model for the Titanic data, obtained by using the `plot()` method in Python.

We can plot AL and LD profiles in a single chart. Toward this end, in the code that follows, we pass the `ld_rf` object, which contains LD profiles, as the first argument of the `plot()` method of the `al_rf` object that includes AL profiles. We also use the `variables` argument to display the profiles only for *age* and *fare*. The resulting plot is presented in Figure 18.9.

```
al_rf.plot(ld_rf, variables = ['age', 'fare'])
```

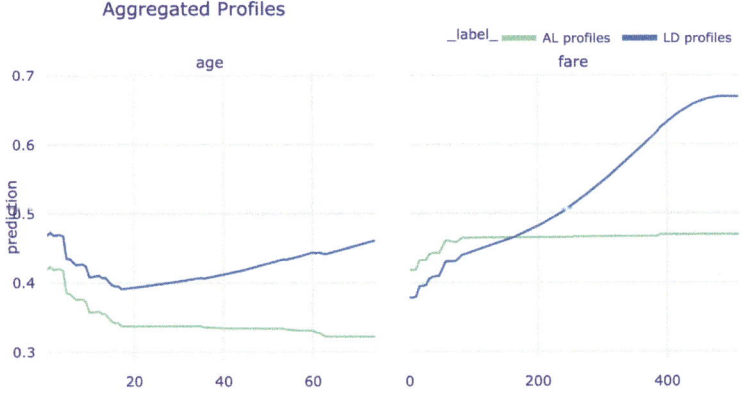

FIGURE 18.9 Local-dependence and accumulated-local profiles for *age* and *fare* for the random forest model for the Titanic data, obtained by using the `plot()` method in Python.

19

Residual-diagnostics Plots

19.1 Introduction

In this chapter, we present methods that are useful for a detailed examination of both overall and instance-specific model performance. In particular, we focus on graphical methods that use residuals. The methods may be used for several purposes:

- In Part II of the book, we discussed tools for single-instance exploration. Residuals can be used to identify potentially problematic instances. The single-instance explainers can then be used in the problematic cases to understand, for instance, which factors contribute most to the errors in prediction.

- For most models, residuals should express a random behavior with certain properties (like, e.g., being concentrated around 0). If we find any systematic deviations from the expected behavior, they may signal an issue with a model (for instance, an omitted explanatory variable or a wrong functional form of a variable included in the model).

- In Chapter 15, we discussed measures that can be used to evaluate the overall performance of a predictive model. Sometimes, however, we may be more interested in cases with the largest prediction errors, which can be identified with the help of residuals.

Residual diagnostics is a classical topic related to statistical modelling. It is most often discussed in the context of the evaluation of goodness-of-fit of a model. That is, residuals are computed using the training data and used to assess whether the model predictions "fit" the observed values of the dependent variable. The literature on the topic is vast, as essentially every book on statistical modeling includes some discussion about residuals. Thus, in this chapter, we are not aiming at being exhaustive. Rather, our goal is to present selected concepts that underlie the use of residuals for predictive models.

19.2   Intuition

As it was mentioned in Section 2.3, we primarily focus on models describing the expected value of the dependent variable as a function of explanatory variables. In such a case, for a "perfect" predictive model, the predicted value of the dependent variable should be exactly equal to the actual value of the variable for every observation. Perfect prediction is rarely, if ever, expected. In practice, we want the predictions to be reasonably close to the actual values. This suggests that we can use the difference between the predicted and the actual value of the dependent variable to quantify the quality of predictions obtained from a model. The difference is called a *residual*.

For a single observation, residual will almost always be different from zero. While a large (absolute) value of a residual may indicate a problem with a prediction for a particular observation, it does not mean that the quality of predictions obtained from a model is unsatisfactory in general. To evaluate the quality, we should investigate the "behavior" of residuals for a group of observations. In other words, we should look at the distribution of the values of residuals.

For a "good" model, residuals should deviate from zero randomly, i.e., not systematically. Thus, their distribution should be symmetric around zero, implying that their mean (or median) value should be zero. Also, residuals should be close to zero themselves, i.e., they should show low variability.

Usually, to verify these properties, graphical methods are used. For instance, a histogram can be used to check the symmetry and location of the distribution of residuals. Note that a model may imply a concrete distribution for residuals. In such a case, the distributional assumption can be verified by using a suitable graphical method like, for instance, a quantile-quantile plot. If the assumption is found to be violated, one might want to be careful when using predictions obtained from the model.

19.3   Method

As it was already mentioned in Chapter 2, for a continuous dependent variable Y, residual r_i for the i-th observation in a dataset is the difference between the observed value of Y and the corresponding model prediction:

$$r_i = y_i - f(\underline{x}_i) = y_i - \widehat{y}_i. \tag{19.1}$$

Standardized residuals are defined as

$$\tilde{r}_i = \frac{r_i}{\sqrt{\text{Var}(r_i)}}, \tag{19.2}$$

where $\text{Var}(r_i)$ is the variance of the residual r_i.

Of course, in practice, the variance of r_i is usually unknown. Hence, the estimated value of $\text{Var}(r_i)$ is used in (19.2). Residuals defined in this way are often called the *Pearson residuals* (Galecki and Burzykowski, 2013). Their distribution should be approximately standard-normal. For the classical linear-regression model, $\text{Var}(r_i)$ can be estimated by using the design matrix. On the other hand, for count data, the variance can be estimated by $f(\underline{x}_i)$, i.e., the expected value of the count. In general, for complicated models, it may be hard to estimate $\text{Var}(r_i)$, so it is often approximated by a constant for all residuals.

Definition (19.2) can also be applied to a binary dependent variable if the model prediction $f(\underline{x}_i)$ is the probability of observing y_i and upon coding the two possible values of the variable as 0 and 1. However, in this case, the range of possible values of r_i is restricted to $[-1, 1]$, which limits the usefulness of the residuals. For this reason, more often the Pearson residuals are used. Note that, if the observed values of the explanatory-variable vectors $\underline{x}_i$ lead to different predictions $f(\underline{x}_i)$ for different observations in a dataset, the distribution of the Pearson residuals will not be approximated by the standard-normal one. This is the case when, for instance, one (or more) of the explanatory variables is continuous. Nevertheless, in that case, the index plot may still be useful to detect observations with large residuals. The standard-normal approximation is more likely to apply in the situation when the observed values of vectors $\underline{x}_i$ split the data into a few, say K, groups, with observations in group k ($k = 1, \ldots, K$) sharing the same predicted value f_k. This may be happen if all explanatory variables are categorical with a limited number of categories. In that case, one can consider averaging residuals r_i per group and standardizing them by $\sqrt{f_k(1 - f_k)/n_k}$, where n_k is the number of observations in group k.

For categorical data, residuals are usually defined in terms of differences in predictions for the dummy binary variable indicating the category observed for the i-th observation.

Let us consider the classical linear-regression model. In that case, residuals should be normally distributed with mean zero and variance defined by the diagonal of hat-matrix $\underline{X}(\underline{X}^T\underline{X})^{-1}\underline{X}^T$. For independent explanatory variables, it should lead to a constant variance of residuals. Figure 19.1 presents examples of classical diagnostic plots for linear-regression models that can be used to check whether the assumptions are fulfilled. In fact, the plots in Figure 19.1 suggest issues with the assumptions.

In particular, the top-left panel presents the residuals in function of the estimated linear combination of explanatory variables, i.e., predicted (fitted) values. For a well-fitting model, the plot should show points scattered symmetrically around the horizontal straight line at 0. However, the scatter in the top-left panel of Figure 19.1 has got a shape of a funnel, reflecting increasing variability of residuals for increasing fitted values. This indicates a violation of the homoscedasticity, i.e., the constancy of variance, assumption. Also, the smoothed line suggests that the mean of residuals becomes increasingly positive for increasing fitted values. This indicates a violation of the assumption that residuals have got zero-mean.

The top-right panel of Figure 19.1 presents the scale-location plot, i.e., the plot of $\sqrt{\tilde{r}_i}$ in function of the fitted values $f(\underline{x}_i)$. For a well-fitting model, the plot should show points scattered symmetrically across the horizontal axis. This is clearly not the case of the plot in Figure 19.1, which indicates a violation of the homoscedasticity assumption.

The bottom-left panel of Figure 19.1 presents the plot of standardized residuals in the function of *leverage*. Leverage is a measure of the distance between $\underline{x}_i$ and the vector of mean values for all explanatory variables (Kutner et al., 2005). A large leverage value for the i-th observation, say l_i, indicates that $\underline{x}_i$ is distant from the center of all observed values of the vector of explanatory variables. Importantly, a large leverage value implies that the observation may have an important influence on predicted/fitted values. In fact, for the classical linear-regression model, it can be shown that the predicted sum-of-squares, defined in (15.5), can be written as

$$PRESS = \sum_{i=1}^{n}(\widehat{y}_{i(-i)} - y_i)^2 = \sum_{i=1}^{n}\frac{r_i^2}{(1-l_i)^2}. \qquad (19.3)$$

Thus, (19.3) indicates that observations with a large r_i (or $\tilde{r}_i$) and a large l_i have an important influence on the overall predictive performance of the model. Hence, the plot of standardized residuals in the function of leverage can be used to detect such influential observations. Note that the plot can also be used to check homoscedasticity because, under that assumption, it should show a symmetric scatter of points around the horizontal line at 0. This is not the case of the plot presented in the bottom-left panel of Figure 19.1. Hence, the plot suggests that the assumption is not fulfilled. However, it does not indicate any particular influential observations, which should be located in the upper-right or lower-right corners of the plot.

Note that the plot of standardized residuals in function of leverage can also be used to detect observations with large differences between the predicted and observed value of the dependent variable. In particular, given that $\tilde{r}_i$ should have approximately standard-normal distribution, only about 0.5% of them should be larger, in absolute value, than 2.57. If there is an excess of such observations, this could be taken as a signal of issues with the fit of the model. At least two such observations (59 and 143) are indicated in the plot shown in the bottom-left panel of Figure 19.1.

Finally, the bottom-right panel of Figure 19.1 presents an example of a normal quantile-quantile plot. In particular, the vertical axis represents the ordered values of the standardized residuals, whereas the horizontal axis represents the corresponding values expected from the standard normal distribution. If the normality assumption is fulfilled, the plot should show a scatter of points close to the 45° diagonal. Clearly, this is not the case of the plot in the bottom-right panel of Figure 19.1.

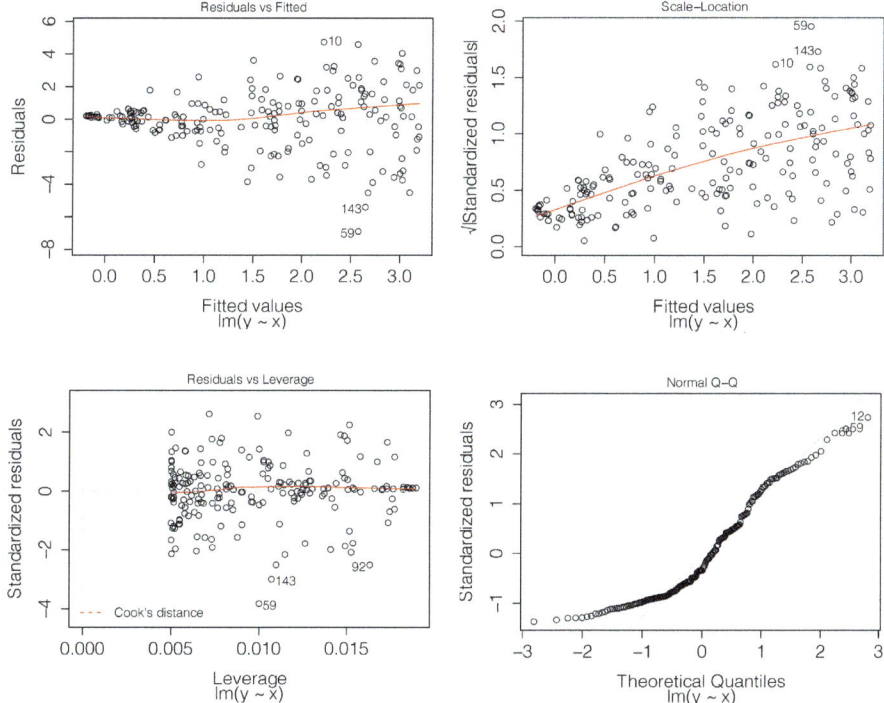

FIGURE 19.1 Diagnostic plots for a linear-regression model. Clockwise from the top-left: residuals in function of fitted values, a scale-location plot, a normal quantile-quantile plot, and a leverage plot. In each panel, indexes of the three most extreme observations are indicated.

19.4   Example: apartment-prices data

In this section, we consider the linear-regression model `apartments_lm` (Section 4.5.1) and the random forest model `apartments_rf` (Section 4.5.2) for the apartment-prices dataset (Section 4.4). Recall that the dependent variable of interest, the price per square meter, is continuous. Thus, we can use residuals r_i, as defined in (19.1). We compute the residuals for the `apartments_test` testing dataset (see Section 4.5.4). It is worth noting that, as it was mentioned in Section 15.4.1, RMSE for both models is very similar for that dataset. Thus, overall, the two models could be seen as performing similarly on average.

Figures 19.2 and 19.3 summarize the distribution of residuals for both models. In particular, Figure 19.2 presents histograms of residuals, while Figure 19.3 shows box-and-whisker plots for the absolute value of the residuals.

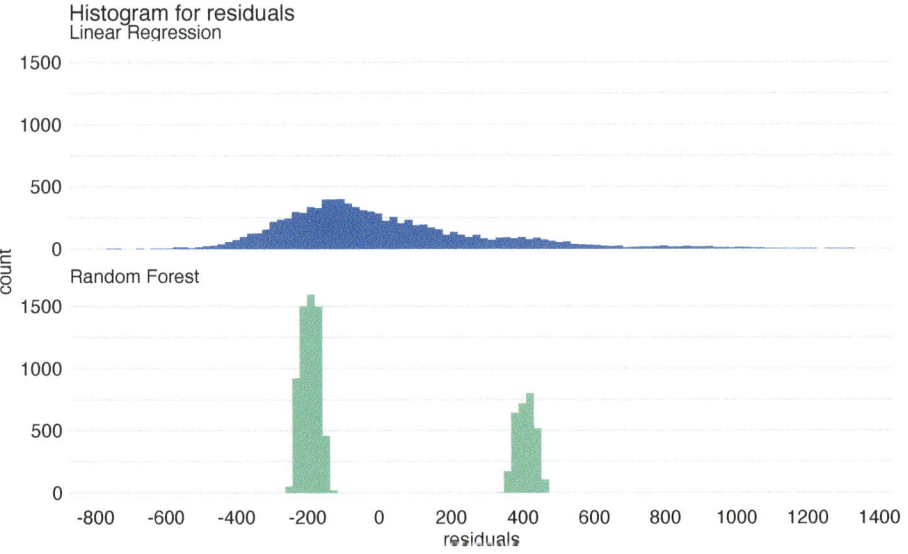

FIGURE 19.2 Histogram of residuals for the linear-regression model `apartments_lm` and the random forest model `apartments_rf` for the `apartments_test` dataset.

Despite the similar value of RMSE, the distributions of residuals for both models are different. In particular, Figure 19.2 indicates that the distribution for the linear-regression model is, in fact, split into two separate, normal-like parts, which may suggest omission of a binary explanatory variable in the model. The two components are located around the values of about -200 and 400. As mentioned in the previous chapters, the reason for this behavior of the residuals is the fact that the model does not capture the non-linear relationship between the price and the year of construction. For instance, Figure 17.8 indicates that the relationship between the construction year and the price may be U-shaped. In particular, apartments built between 1940 and 1990 appear to be, on average, cheaper than those built earlier or later.

As seen from Figure 19.2, the distribution of residuals for the random forest model is skewed to the right and multimodal. It seems to be centered at a value closer to zero than the distribution for the linear-regression model, but it shows a larger variation. These conclusions are confirmed by the box-and-whisker plots in Figure 19.3.

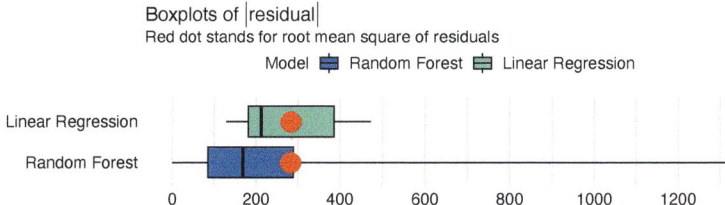

FIGURE 19.3 Box-and-whisker plots of the absolute values of the residuals of the linear-regression model `apartments_lm` and the random forest model `apartments_rf` for the `apartments_test` dataset. The dots indicate the mean value that corresponds to root-mean-squared-error.

The plots in Figures 19.2 and 19.3 suggest that the residuals for the random forest model are more frequently smaller than the residuals for the linear-regression model. However, a small fraction of the random forest-model residuals is very large, and it is due to them that the RMSE is comparable for the two models.

In the remainder of the section, we focus on the random forest model.

Figure 19.4 shows a scatter plot of residuals (vertical axis) in function of the observed (horizontal axis) values of the dependent variable. For a "perfect" predictive model, we would expect the horizontal line at zero. For a "good" model, we would like to see a symmetric scatter of points around the horizontal line at zero, indicating random deviations of predictions from the observed values. The plot in Figure 19.4 shows that, for the large observed values of the dependent variable, the residuals are positive, while for small values they are negative. This trend is clearly captured by the smoothed curve included in the graph. Thus, the plot suggests that the predictions are shifted (biased) towards the average.

The shift towards the average can also be seen from Figure 19.5 that shows a scatter plot of the predicted (vertical axis) and observed (horizontal axis) values of the dependent variable. For a "perfectly" fitting model we would expect a diagonal line (indicated in red). The plot shows that, for large observed values of the dependent variable, the predictions are smaller than the observed

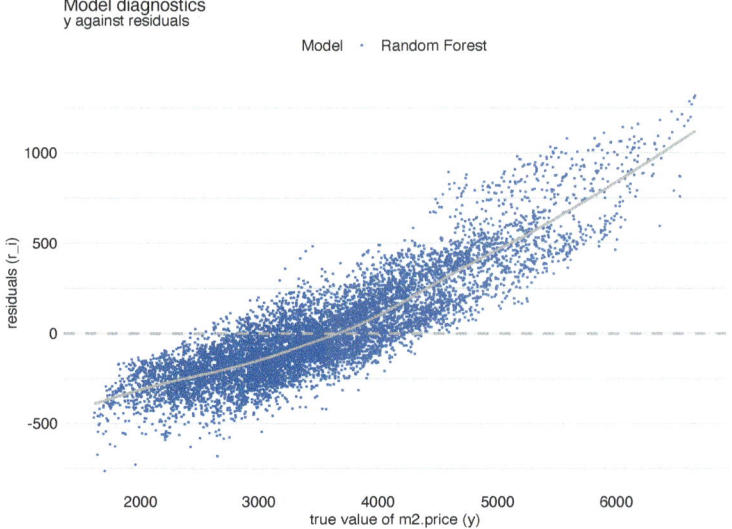

FIGURE 19.4 Residuals and observed values of the dependent variable for the random forest model `apartments_rf` for the `apartments_test` dataset.

values, with an opposite trend for the small observed values of the dependent variable.

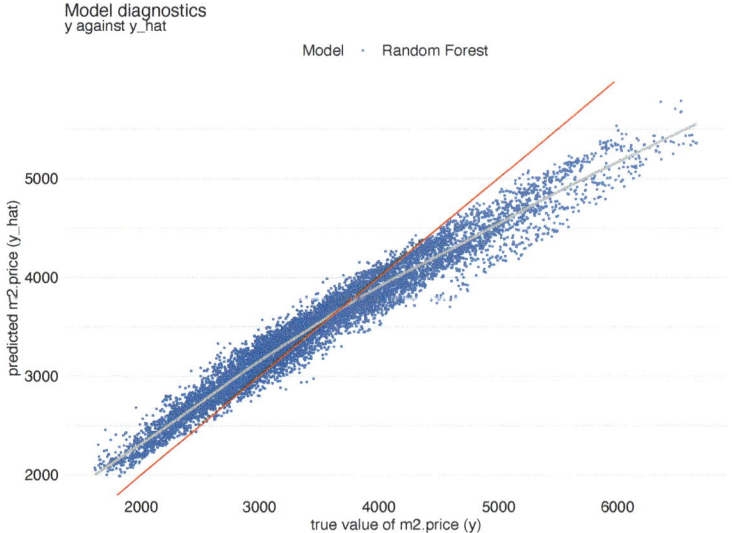

FIGURE 19.5 Predicted and observed values of the dependent variable for the random forest model `apartments_rf` for the `apartments_test` dataset. The red line indicates the diagonal.

Figure 19.6 shows an index plot of residuals, i.e., their scatter plot in function of an (arbitrary) identifier of the observation (horizontal axis). The plot indicates an asymmetric distribution of residuals around zero, as there is an excess of large positive (larger than 500) residuals without a corresponding fraction of negative values. This can be linked to the right-skewed distribution seen in Figures 19.2 and 19.3 for the random forest model.

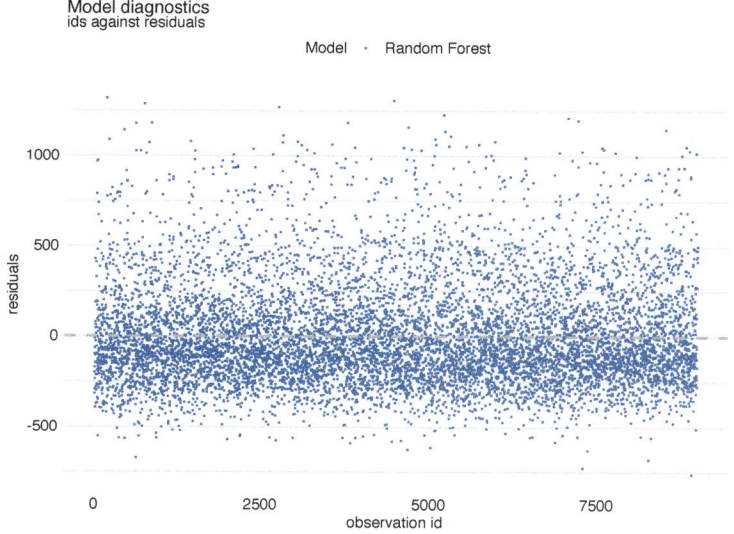

FIGURE 19.6 Index plot of residuals for the random forest model `apartments_rf` for the `apartments_test` dataset.

Figure 19.7 shows a scatter plot of residuals (vertical axis) in function of the predicted (horizontal axis) value of the dependent variable. For a "good" model, we would like to see a symmetric scatter of points around the horizontal line at zero. The plot in Figure 19.7, as the one in Figure 19.4, suggests that the predictions are shifted (biased) towards the average.

The random forest model, as the linear-regression model, assumes that residuals should be homoscedastic, i.e., that they should have a constant variance. Figure 19.8 presents a variant of the scale-location plot of residuals, i.e., a scatter plot of the absolute value of residuals (vertical axis) in function of the predicted values of the dependent variable (horizontal axis). The plot includes a smoothed line capturing the average trend. For homoscedastic residuals, we would expect a symmetric scatter around a horizontal line; the smoothed trend should be also horizontal. The plot in Figure 19.8 deviates from the expected pattern and indicates that the variability of the residuals depends on the (predicted) value of the dependent variable.

For models like linear regression, such heteroscedasticity of the residuals would be worrying. In random forest models, however, it may be less of concern. This

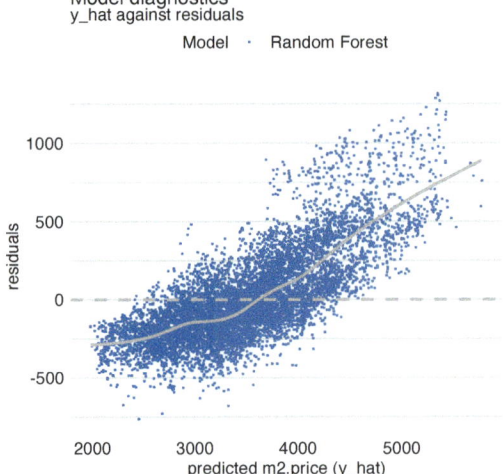

FIGURE 19.7 Residuals and predicted values of the dependent variable for the random forest model `apartments_rf` for the `apartments_test` dataset.

is beacuse it may occur due to the fact that the models reduce variability of residuals by introducing a bias (towards the average). Thus, it is up to the developer of a model to decide whether such a bias (in our example, for the cheapest and most expensive apartments) is a desirable price to pay for the reduced residual variability.

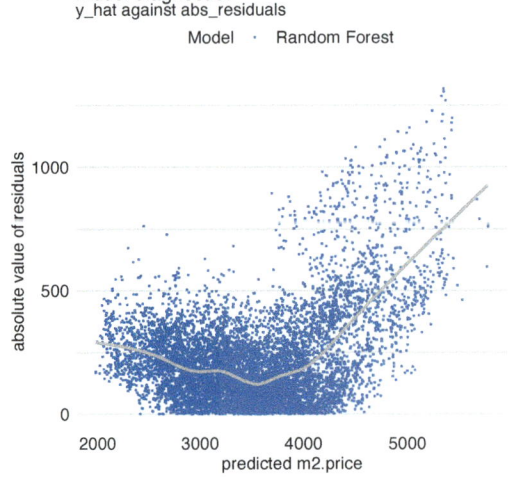

FIGURE 19.8 The scale-location plot of residuals for the random forest model `apartments_rf` for the `apartments_test` dataset.

19.5 Pros and cons

Diagnostic methods based on residuals are a very useful tool in model exploration. They allow identifying different types of issues with model fit or prediction, such as problems with distributional assumptions or with the assumed structure of the model (in terms of the selection of the explanatory variables and their form). The methods can help in detecting groups of observations for which a model's predictions are biased and, hence, require inspection.

A potential complication related to the use of residual diagnostics is that they rely on graphical displays. Hence, for a proper evaluation of a model, one may have to construct and review many graphs. Moreover, interpretation of the patterns seen in graphs may not be straightforward. Also, it may not be immediately obvious which element of the model may have to be changed to remove the potential issue with the model fit or predictions.

19.6 Code snippets for R

In this section, we present diagnostic plots as implemented in the `DALEX` package for R. The package covers all plots and methods presented in this chapter. Similar functions can be found in packages `auditor` (Gosiewska and Biecek, 2018), `rms` (Harrell Jr, 2018), and `stats` (Faraway, 2005).

For illustration purposes, we will show how to create the plots shown in Section 19.4 for the linear-regression model `apartments_lm` (Section 4.5.1) and the random forest model `apartments_rf` (Section 4.5.2) for the `apartments_test` dataset (Section 4.4).

We first load the two models via the `archivist` hooks, as listed in Section 4.5.6. Subsequently, we construct the corresponding explainers by using function `explain()` from the `DALEX` package (see Section 4.2.6). Note that we use the `apartments_test` data frame without the first column, i.e., the *m2.price* variable, in the `data` argument. This will be the dataset to which the model will be applied. The *m2.price* variable is explicitly specified as the dependent variable in the `y` argument. We also load the `randomForest` package, as it is important to have the corresponding `predict()` function available for the random forest model.

```
library("DALEX")
model_apart_lm <- archivist:: aread("pbiecek/models/55f19")
explain_apart_lm <- DALEX::explain(model = model_apart_lm,
                         data    = apartments_test[,-1],
                         y       = apartments_test$m2.price,
                         label   = "Linear Regression")
library("randomForest")
model_apart_rf <- archivist:: aread("pbiecek/models/fe7a5")
explain_apart_rf <- DALEX::explain(model = model_apart_rf,
                         data    = apartments_test[,-1],
                         y       = apartments_test$m2.price,
                         label   = "Random Forest")
```

For exploration of residuals, DALEX includes two useful functions. The model_performance() function can be used to evaluate the distribution of the residuals. On the other hand, the model_diagnostics() function is suitable for investigating the relationship between residuals and other variables.

The model_performance() function was already introduced in Section 15.6. Application of the function to an explainer-object returns an object of class "model_performance" which includes, in addition to selected model-performance measures, a data frame containing the observed and predicted values of the dependent variable together with the residuals.

```
mr_lm <- DALEX::model_performance(explain_apart_lm)
mr_rf <- DALEX::model_performance(explain_apart_rf)
```

By applying the plot() function to a "model_performance"-class object we can obtain various plots. The required type of the plot is specified with the help of the geom argument (see Section 15.6). In particular, specifying geom = "histogram" results in a histogram of residuals. In the code below, we apply the plot() function to the "model_performance"-class objects for the linear-regression and random forest models. As a result, we automatically get a single graph with the histograms of residuals for the two models. The resulting graph is shown in Figure 19.2

```
library("ggplot2")
plot(mr_lm, mr_rf, geom = "histogram")
```

The box-and-whisker plots of the residuals for the two models can be constructed by applying the geom = "boxplot" argument. The resulting graph is shown in Figure 19.3.

```
plot(mr_lm, mr_rf, geom = "boxplot")
```

Function `model_diagnostics()` can be applied to an explainer-object to directly compute residuals. The resulting object of class "model_diagnostics" is a data frame in which the residuals and their absolute values are combined with the observed and predicted values of the dependent variable and the observed values of the explanatory variables. The data frame can be used to create various plots illustrating the relationship between residuals and the other variables.

```
md_lm <- model_diagnostics(explain_apart_lm)
md_rf <- model_diagnostics(explain_apart_rf)
```

Application of the `plot()` function to a `model_diagnostics`-class object produces, by default, a scatter plot of residuals (on the vertical axis) in function of the predicted values of the dependent variable (on the horizontal axis). By using arguments `variable` and `yvariable`, it is possible to specify plots with other variables used for the horizontal and vertical axes, respectively. The two arguments accept, apart from the names of the explanatory variables, the following values:

- `"y"` for the dependent variable,
- `"y_hat"` for the predicted value of the dependent variable,
- `"obs"` for the identifiers of observations,
- `"residuals"` for residuals,
- `"abs_residuals"` for absolute values of residuals.

Thus, to obtain the plot of residuals in function of the observed values of the dependent variable, as shown in Figure 19.4, the syntax presented below can be used.

```
plot(md_rf, variable = "y", yvariable = "residuals")
```

To produce Figure 19.5, we have got to use the predicted values of the dependent variable on the vertical axis. This is achieved by specifying the `yvariable = "y_hat"` argument. We add the diagonal reference line to the plot by using the `geom_abline()` function.

```
plot(md_rf, variable = "y", yvariable = "y_hat") +
    geom_abline(colour = "red", intercept = 0, slope = 1)
```

Figure 19.6 presents an index plot of residuals, i.e., residuals (on the vertical axis) in function of identifiers of individual observations (on the horizontal axis). Toward this aim, we use the `plot()` function call as below.

```
plot(md_rf, variable = "ids", yvariable = "residuals")
```

Finally, Figure 19.8 presents a variant of the scale-location plot, with absolute values of the residuals shown on the vertical scale and the predicted values of the dependent variable on the horizontal scale. The plot is obtained with the syntax shown below.

```
plot(md_rf, variable = "y_hat", yvariable = "abs_residuals")
```

Note that, by default, all plots produced by applying the `plot()` function to a "model_diagnostics"-class object include a smoothed curve. To exclude the curve from a plot, one can use the argument `smooth = FALSE`.

19.7   Code snippets for Python

In this section, we use the `dalex` library for Python. The package covers all methods presented in this chapter. But, as mentioned in Section 19.1, residuals are a classical model-diagnostics tool. Thus, essentially any model-related library includes functions that allow calculation and plotting of residuals.

For illustration purposes, we use the `apartments_rf` random forest model for the Titanic data developed in Section 4.6.2. Recall that the model is developed to predict the price per square meter of an apartment in Warsaw.

In the first step, we create an explainer-object that will provide a uniform interface for the predictive model. We use the `Explainer()` constructor for this purpose.

```
import dalex as dx
apartments_rf_exp = dx.Explainer(apartments_rf, X, y,
      label = "Apartments RF Pipeline")
```

The function that calculates residuals, absolute residuals and observation ids is `model_diagnostics()`.

```
md_rf = apartments_rf_exp.model_diagnostics()
md_rf.result
```

	construction_year	surface	floor	no_rooms	district	y	y_hat	residuals	abs_residuals	label	ids
1	1953	25	3	1	Srodmiescie	5897	5745.522985	151.477015	151.477015	Apartments RF Pipeline	1
2	1992	143	9	5	Bielany	1818	2735.719742	-917.719742	917.719742	Apartments RF Pipeline	2
3	1937	56	1	2	Praga	3643	3903.124847	-260.124847	260.124847	Apartments RF Pipeline	3
4	1995	93	7	3	Ochota	3517	2902.279790	614.720210	614.720210	Apartments RF Pipeline	4
5	1992	144	6	5	Mokotow	3013	3034.865606	-21.865606	21.865606	Apartments RF Pipeline	5
...	...	...	...	...	...	...	...	...	...	...	...
996	1921	44	2	2	Srodmiescie	6355	6040.295564	314.704436	314.704436	Apartments RF Pipeline	996
997	1921	48	10	2	Bemowo	3422	3599.223815	-177.223815	177.223815	Apartments RF Pipeline	997
998	1980	85	3	3	Bemowo	3098	3157.624542	-59.624542	59.624542	Apartments RF Pipeline	998
999	1942	36	7	1	Zoliborz	4192	3606.504654	585.495346	585.495346	Apartments RF Pipeline	999
1000	1992	112	6	5	Mokotow	3327	3049.726695	277.273305	277.273305	Apartments RF Pipeline	1000

The results can be visualised by applying the `plot()` method. Figure 19.9 presents the created plot.

```
md_rf.plot()
```

In the `plot()` function, we can specify what shall be presented on horizontal

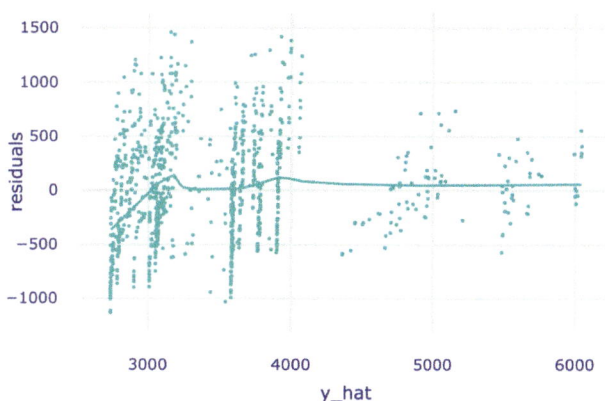

FIGURE 19.9 Residuals versus predicted values for the random forest model for the Apartments data.

and vertical axes. Possible values are columns in the `md_rf.result` data frame, i.e. `residuals`, `abs_residuals`, `y`, `y_hat`, `ids` and variable names.

```
mp_rf.plot(variable = "ids", yvariable = "abs_residuals")
```

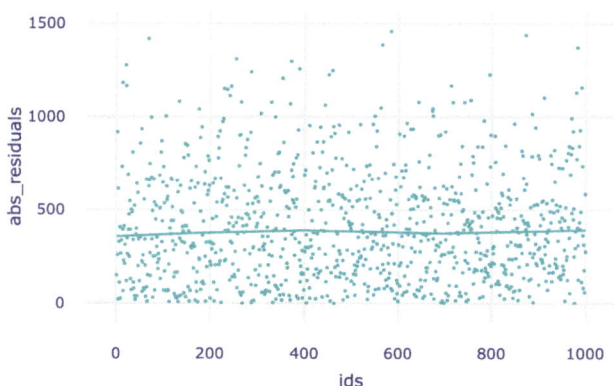

FIGURE 19.10 Absolute residuals versus indices of corresponding observations for the random forest model for the Apartments data.

20

Summary of Dataset-level Exploration

20.1 Introduction

In Part III of the book, we introduced several techniques for global exploration and explanation of a model's predictions for a set of instances. Each chapter was devoted to a single technique. In practice, these techniques are rarely used separately. Rather, it is more informative to combine different insights offered by each technique into a more holistic overview.

Figure 20.1 offers a graphical illustration of the idea. The graph includes the results of different dataset-level explanation techniques applied to the random forest model (Section 4.2.2) for the Titanic data (Section 4.1).

The plots in the first row of Figure 20.1 show how good is the model and which variables are the most important. In particular, the first two graphs in that row present measures of the model's overall performance, as introduced in Chapter 15, along with a graphical summary in the form of the ROC curve. The last plot in the row shows values of the variable-importance measure obtained by the method introduced in Chapter 16. The three graphs indicate a reasonable overall performance of the model and suggest that the most important variables are *gender*, *class*, and *age*.

The two first plots in the second row of Figure 20.1 show partial-dependence (PD) profiles for *age* (continuous variable) and *class* (categorical variable). The profile for *age* suggest that the critical cutoff, from a perspective of the variable's effect, is around 18 years. The profile for *class* indicates that, among different travel-classes, the largest effect and chance for survival is associated with the deck crew. The last graph on the right in the second row of Figure 20.1 presents a scatter plot of residuals for *age*. In the plot, residuals for passengers with age between 20 and 40 years are, on average, slightly larger than for other ages, but the bias is not excessive.

The plots in the third row of Figure 20.1 summarize distributions of (from the left to the right the) *age*, *fare*, *gender*, and *class*.

Thus, Figure 20.1 illustrates that perspectives offered by different techniques complement each other and, when combined, allow obtaining a more profound insight into the model's predictive performance.

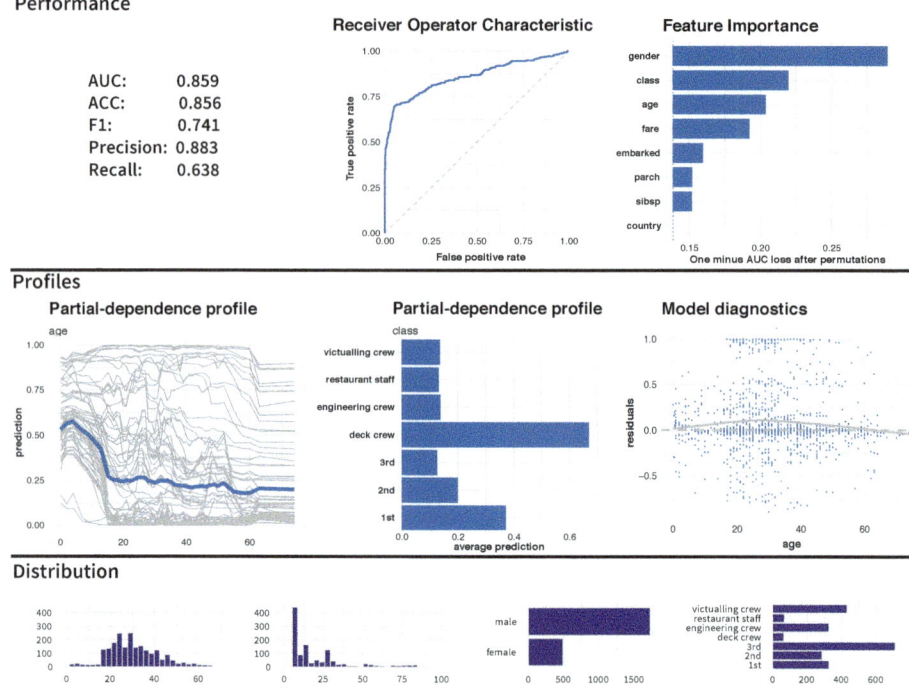

FIGURE 20.1 Results of dataset-level explanation techniques for the random forest model for the Titanic data.

Figure 20.2 illustrates how the information obtained from model exploration can be used to gain a better understanding of the model and of the domain that is being modelled. Modelling is an iterative process (see Section 2.2), and the more we learn in each iteration, the easier it is to plan and execute the next one.

While combining various techniques for dataset-level explanation can provide additional insights, it is worth remembering that the techniques are, indeed, different and their suitability may depend on the problem at hand. This is what we discuss in the remainder of the chapter.

20.2   Exploration on training/testing data

The key element of dataset-level explanatory techniques is the set of observations on which the exploration is performed. Therefore, sometimes these

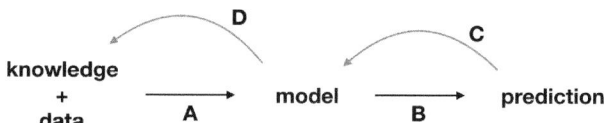

FIGURE 20.2 Explainability techniques allow strengthening the feedback extracted from a model. A, data and domain knowledge allow building the model. B, predictions are obtained from the model. C, by analyzing the predictions, we learn more about the model. D, better understanding of the model allows better understanding of the data and, sometimes, broadens domain knowledge.

techniques are called dataset-level exploration. This is the terminology that we have adopted.

In the case of machine-learning models, data are often split into a training set and a testing set. Which one should we use for model exploration?

In the case of model-performance assessment, it is natural to evaluate it using an independent testing dataset to minimize the risk of overfitting. However, PD profiles or accumulated-local (AL) profiles can be constructed for both the training and testing datasets. If the model is generalizable, its behavior for the two datasets should be similar. If we notice significant differences in the results for the two datasets, the source of the differences should be examined, as it may be a sign of shifts in variable distributions. Such shifts are called data-drift in the machine learning world.

20.3 Correlated explanatory variables

Most methods introduced in Part III of the book analyze each exploratory variable independently from the other variables. However, more often than not, exploratory variables are correlated and jointly present some aspects of observations. For example, *fare* and *class* in the Titanic data are correlated (see Section 4.1.1), and both are related to the wealth of the passenger. Many of the techniques presented in Part III can be generalized to allow a joint analysis of groups of two or more exploratory variables. For example, the permutation-based variable-importance measure presented in Chapter 16 was introduced in the context of an analysis of a single variable. However, permutations can be done for a group of variables, allowing for evaluation of the importance of the entire group.

Similarly, PD profiles (Chapter 17) assumed independence of variables. However, AL profiles (Chapter 18) allow taking correlation between variables into account. Moreover, it is possible to extend the AL profiles to the analysis of interaction effects.

20.4   Comparison of models (champion-challenger analysis)

One of important applications of the methods presented in Part III of the book is comparison of models.

It appears that different models may offer a similar performance, while basing their predictions on different relations extracted from the same data. Breiman (2001b) refers to this phenomenon as "Rashomon effect". By comparing models, we can gain important insights that can lead to improvement of one or several of the models. For instance, when comparing a more flexible model with a more rigid one, we can check if they discover similar relationships between the dependent variable and explanatory variables. If they do, this can reassure us that both models have discovered genuine aspects of the data. Sometimes, however, the rigid model may miss a relationship that might have been found by the more flexible one. This could provide, for instance, a suggestion for an improvement of the former. An example of such a case was provided in Chapter 17, in which a linear-regression model was compared to a random forest model for the apartment-prices dataset. The random forest model suggested a U-shaped relationship between the construction year and the price of an apartment, which was missed by the simple linear-regression model.

Part IV

Use-cases

21

FIFA 19

21.1 Introduction

In the previous chapters, we introduced a range of methods for the exploration of predictive models. Different methods were discussed in separate chapters, and while illustrated, they were not directly compared. Thus, in this chapter, we apply the methods to one dataset in order to present their relative merits. In particular, we present an example of a full process of a model development along the lines introduced in Chapter 2. This will allow us to show how one can combine results from different methods.

The Fédération Internationale de Football Association (FIFA) is a governing body of football (sometimes, especially in the USA, called soccer). FIFA is also a series of video games developed by EA Sports which faithfully reproduces the characteristics of real players. FIFA ratings of football players from the video game can be found at `https://sofifa.com/`. Data from this website for 2019 were scrapped and made available at the Kaggle webpage `https://www.kaggle.com/karangadiya/fifa19`.

We will use the data to build a predictive model for the evaluation of a player's value. Subsequently, we will use the model exploration and explanation methods to better understand the model's performance, as well as which variables and how to influence a player's value.

21.2 Data preparation

The original dataset contains 89 variables that describe 16,924 players. The variables include information such as age, nationality, club, wage, etc. In what follows, we focus on 45 variables that are included in data frame `fifa` included in the `DALEX` package for R and Python. The variables from this dataset set are listed in Table 21.1.

TABLE 21.1: Variables in the FIFA 19 dataset.

Name	Weak.Foot	FKAccuracy	Jumping	Composure
Club	Skill.Moves	LongPassing	Stamina	Marking
Position	Crossing	BallControl	Strength	StandingTackle
Value.EUR	Finishing	Acceleration	LongShots	SlidingTackle
Age	HeadingAccuracy	SprintSpeed	Aggression	GKDiving
Overall	ShortPassing	Agility	Interceptions	GKHandling
Special	Volleys	Reactions	Positioning	GKKicking
Preferred.Foot	Dribbling	Balance	Vision	GKPositioning
Reputation	Curve	ShotPower	Penalties	GKReflexes

In particular, variable `Value.EUR` contains the player's value in millions of EUR. This will be our dependent variable.

The distribution of the variable is heavily skewed to the right. In particular, the quartiles are equal to 325,000 EUR, 725,000 EUR, and 2,534,478 EUR. There are three players with a value higher than 100 millions of Euro.

Thus, in our analyses, we will consider a logarithmically-transformed players' value. Figure 21.1 presents the empirical cumulative-distribution function and histogram for the transformed value. They indicate that the transformation makes the distribution less skewed.

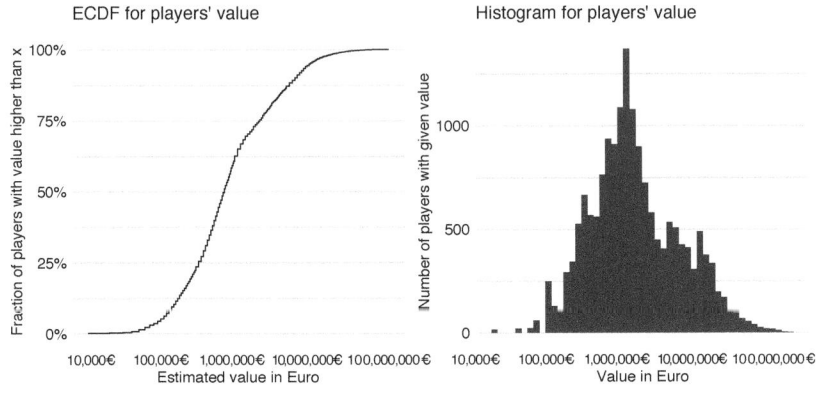

FIGURE 21.1 The empirical cumulative-distribution function and histogram for the $\log_{10}$-transformed players' values.

Additionally, we take a closer look at four characteristics that will be considered as explanatory variables later in this chapter. These are: `Age`, `Reactions` (a movement skill), `BallControl` (a general skill), and `Dribbling` (a general skill).

Figure 21.2 presents histograms of the values of the four variables. From the plot for `Age` we can conclude that most of the players are between 20 and 30 years of age (median age: 25). Variable `Reactions` has an approximately symmetric distribution, with quartiles equal to 56, 62, and 68. Histograms of `BallControl` and `Dribbling` indicate, interestingly, bimodal distributions. The smaller modes are due to goalkeepers.

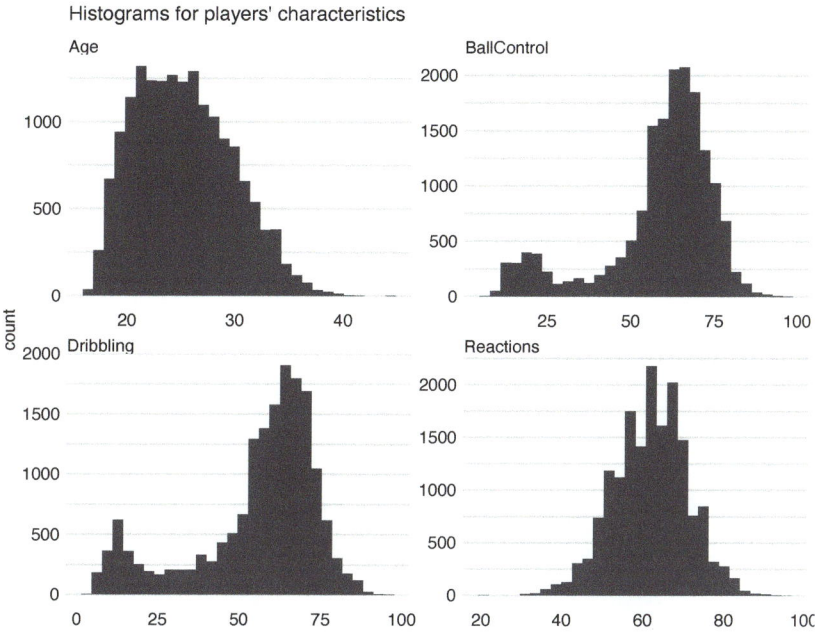

FIGURE 21.2 Histograms for selected characteristics of players.

21.2.1 Code snippets for R

The subset of 5000 most valuable players from the FIFA 19 data is available in the `fifa` data frame in the `DALEX` package.

```
library("DALEX")
head(fifa)
```

21.2.2 Code snippets for Python

The subset of 5000 most valuable players from FIFA 19 data can be loaded to Python with `dalex.datasets.load_fifa()` method.

```
import dalex as dx
fifa = dx.datasets.load_fifa()
```

21.3 Data understanding

We will investigate the relationship between the four selected characteristics and the (logarithmically-transformed) player's value. Toward this aim, we use the scatter plots shown in Figure 21.3. Each plot includes a smoothed curve capturing the trend.

For `Age`, the relationship is not monotonic. There seems to be an optimal age, between 25 and 30 years, at which the player's value reaches the maximum. On the other hand, the value of youngest and oldest players is about 10 times lower, as compared to the maximum.

For variables `BallControl` and `Dribbling`, the relationship is not monotonic. In general, the larger value of these coefficients, the large value of a player. However, there are "local" maxima for players with low scores for `BallControl` and `Dribbling`. As it was suggested earlier, these are probably goalkeepers.

For `Reactions`, the association with the player's value is monotonic, with increasing values of the variable leading to increasing values of players.

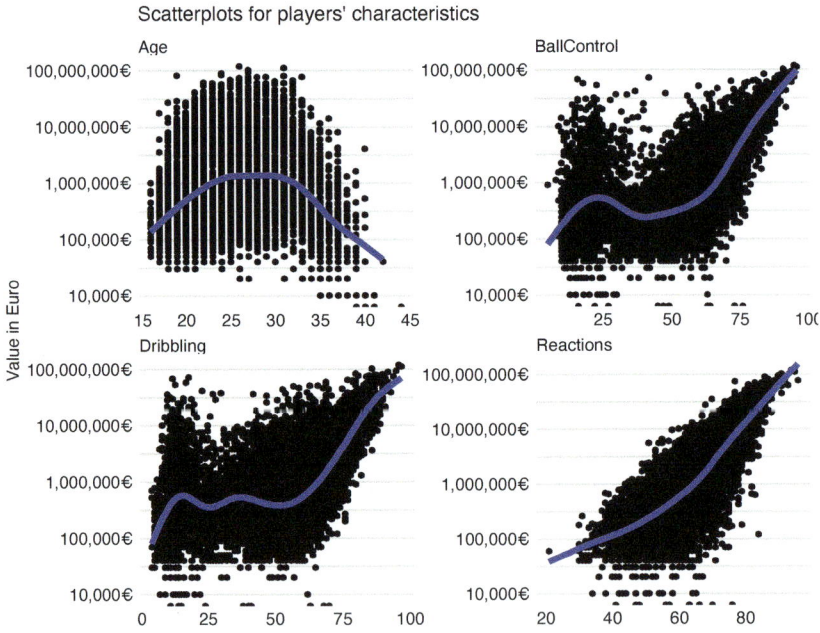

FIGURE 21.3 Scatter plots illustrating the relationship between the (logarithmically-transformed) player's value and selected characteristics.

Figure 21.4 presents the scatter-plot matrix for the four selected variables. It indicates that all variables are positively correlated, though with different strength. In particular, `BallControl` and `Dribbling` are strongly correlated, with the estimated correlation coefficient larger than 0.9. `Reactions` is moderately correlated with the other three variables. Finally, there is a moderate correlation between `Age` and `Reactions`, but not much correlation with `BallControl` and `Dribbling`.

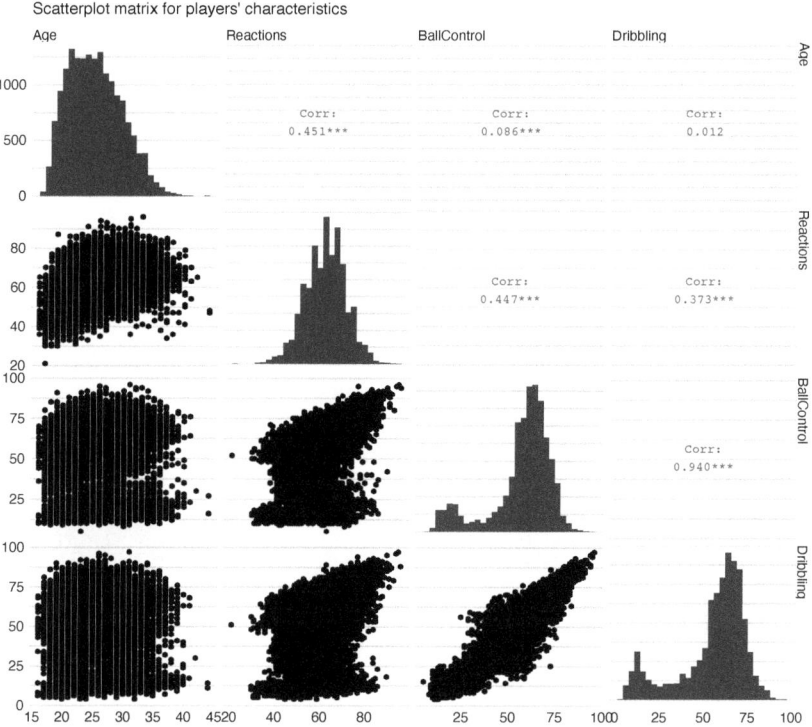

FIGURE 21.4 Scatter-plot matrix illustrating the relationship between selected characteristics of players.

21.4 Model assembly

In this section, we develop a model for players' values. We consider all variables other than `Name`, `Club`, `Position`, `Value.EUR`, `Overall`, and `Special` (see Section 21.2) as explanatory variables. The base-10 logarithm of the player's value is the dependent variable.

Given different possible forms of relationship between the (logarithmically-transformed) player's value and explanatory variables (as seen, for example, in

Figure 21.3), we build four different, flexible models to check whether they are capable of capturing the various relationships. In particular, we consider the following models:

- a boosting model with 250 trees of 1-level depth, as implemented in package **gbm** (Ridgeway, 2017),
- a boosting model with 250 trees of 4-levels depth (this model should be able to catch interactions between variables),
- a random forest model with 250 trees, as implemented in package `ranger` (Wright and Ziegler, 2017),
- a linear model with a spline-transformation of explanatory variables, as implemented in package `rms` (Harrell Jr, 2018).

These models will be explored in detail in the following sections.

21.4.1 Code snippets for R

In this section, we show R-code snippets used to develop the gradient boosting model. Other models were built in a similar way.

The code below fits the model to the data. The dependent variable `LogValue` contains the base-10 logarithm of `Value.EUR`, i.e., of the player's value.

```
fifa$LogValue <- log10(fifa$Value.EUR)
fifa_small <- fifa[,-c(1, 2, 3, 4, 6, 7)]

fifa_gbm_deep <- gbm(LogValue~., data = fifa_small, n.trees = 250,
    interaction.depth = 4, distribution = "gaussian")
```

For model-exploration purposes, we have got to create an explainer-object with the help of the `DALEX::explain()` function (see Section 4.2.6). The code below is used for the gradient boosting model. Note that the model was fitted to the logarithmically-transformed player's value. However, it is more natural to interpret the predictions on the original scale. This is why, in the provided syntax, we apply the `predict_function` argument to specify a user-defined function to obtain predictions on the original scale, in Euro. Additionally, we use the `data` and `y` arguments to indicate the data frame with explanatory variables and the values of the dependent variable, for which predictions are to be obtained. Finally, the model receives its own `label`.

```
library("DALEX")
fifa_gbm_exp_deep <- DALEX::explain(fifa_gbm_deep,
    data = fifa_small, y = 10^fifa_small$LogValue,
    predict_function = function(m,x) 10^predict(m, x, n.trees = 250),
    label = "GBM deep")
```

21.4.2 Code snippets for Python

In this section, we show Python-code snippets used to develop the gradient boosting model. Other models were built in a similar way.

The code below fits the model to the data. The dependent variable `ylog` contains the logarithm of `value_eur`, i.e., of the player's value.

```
from lightgbm import LGBMRegressor
from sklearn.model_selection import train_test_split
import numpy as np

X = fifa.drop(["nationality", "overall", "potential",
    "value_eur", "wage_eur"], axis = 1)
y = fifa['value_eur']
ylog = np.log(y)

X_train, X_test, ylog_train, ylog_test, y_train, y_test =
    train_test_split(X, ylog, y, test_size = 0.25, random_state = 4)
gbm_model = LGBMRegressor()
gbm_model.fit(X_train, ylog_train, verbose = False)
```

For model-exploration purposes, we have to create the explainer-object with the help of the `Explainer()` constructor from the `dalex` library (see Section 4.3.6). The code is provided below. Note that the model was fitted to the logarithmically-transformed player's value. However, it is more natural to interpret the predictions on the original scale. This is why, in the provided syntax, we apply the `predict_function` argument to specify a user-defined function to obtain predictions on the original scale, in Euro. Additionally, we use the `X` and `y` arguments to indicate the data frame with explanatory variables and the values of the dependent variable, for which predictions are to be obtained. Finally, the model receives its own `label`.

```
def predict_function(model, data):
    return np.exp(model.predict(data))

fifa_gbm_exp = dx.Explainer(gbm_model, X_test, y_test,
    predict_function = predict_function, label = 'gbm')
```

21.5 Model audit

Having developed the four candidate models, we may want to evaluate their performance. Toward this aim, we can use the measures discussed in Section 15.3.1. The computed values are presented in Table 21.2. On average, the values of the root-mean-squared-error (RMSE) and mean-absolute-deviation (MAD) are the smallest for the random forest model.

TABLE 21.2: Model-performance measures for the four models for the FIFA 19 data.

	MSE	RMSE	R2	MAD
GBM shallow	8.990×10^{12}	2998449	0.7300429	183682.91
GBM deep	2.211×10^{12}	1487091	0.9335987	118425.56
RF	1.141×10^{12}	1068258	0.9657347	50693.24
RM	21.912×10^{12}	4681129	0.3420350	148187.06

In addition to computing measures of the overall performance of the model, we should conduct a more detailed examination of both overall- and instance-specific performance. Toward this aim, we can apply residual diagnostics, as discussed in Chapter 19.

For instance, we can create a plot comparing the predicted (fitted) and observed values of the dependent variable.

The resulting plot is shown in Figure 21.5. It indicates that predictions are closest to the observed values of the dependent variable for the random forest model. It is worth noting that the smoothed trend for the model is close to a straight line, but with a slope smaller than 1. This implies the random forest model underestimates the actual value of the most expensive players, while it overestimates the value for the least expensive ones. A similar pattern can be observed for the gradient boosting models. This "shrinking to the mean" is typical for this type of models.

21.5.1   Code snippets for R

In this section, we show R-code snippets for model audit for the gradient boosting model. For other models a similar syntax was used.

The `model_performance()` function (see Section 15.6) is used to calculate the values of RMSE, MSE, R^2, and MAD for the model.

```
model_performance(fifa_gbm_exp_deep)
```

The `model_diagnostics()` function (see Section 19.6) is used to create residual-diagnostics plots. Results of this function can be visualised with the generic `plot()` function. In the code that follows, additional arguments are used to improve the outlook and interpretability of both axes.

```
fifa_md_gbm_deep <- model_diagnostics(fifa_gbm_exp_deep)
plot(fifa_md_gbm_deep,
     variable = "y", yvariable = "y_hat") +
```

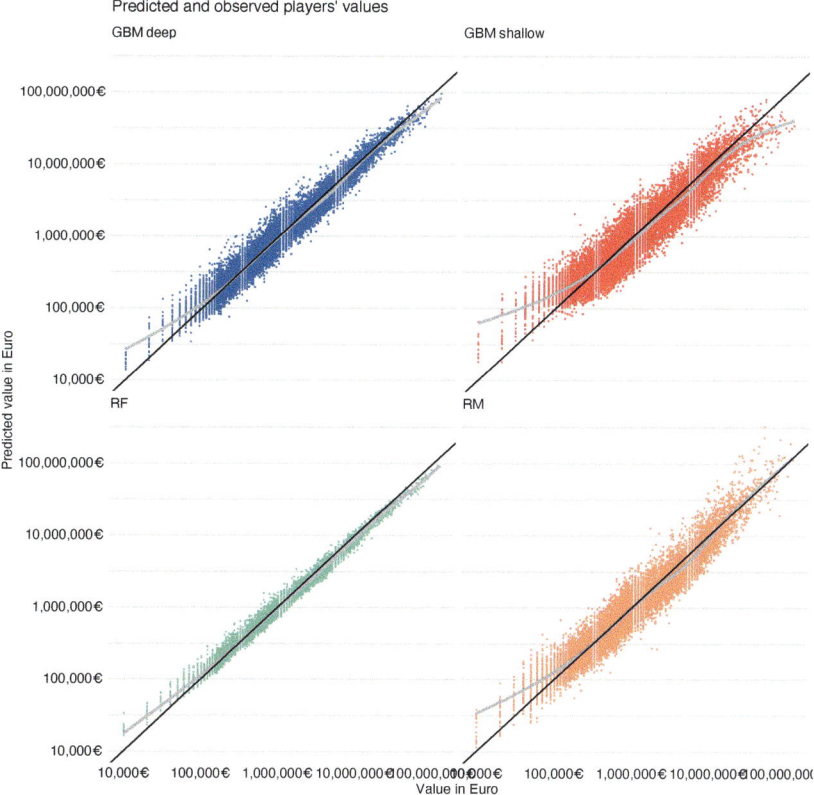

FIGURE 21.5 Observed and predicted (fitted) players' values for the four models for the FIFA 19 data.

```
scale_x_continuous("Value in Euro", trans = "log10",
                labels = dollar_format(suffix = "€", prefix = "")) +
scale_y_continuous("Predicted value in Euro", trans = "log10",
                labels = dollar_format(suffix = "€", prefix = "")) +
geom_abline(slope = 1) +
ggtitle("Predicted and observed players' values", "")
```

21.5.2 Code snippets for Python

In this section, we show Python-code snippets used to perform residual diagnostic for the trained gradient boosting model. Other models were tested in a similar way.

The `fifa_gbm_exp.model_diagnostics()` function (see Section 19.7) is used to calculate the residuals and absolute residuals. Results of this function can be visualised with the `plot()` function. The code below produce diagnostic plots similar to these presented in Figure 21.5.

```
fifa_md_gbm = fifa_gbm_exp.model_diagnostics()
fifa_md_gbm.plot(variable = "y", yvariable = "y_hat")
```

21.6   Model understanding (dataset-level explanations)

All four developed models involve many explanatory variables. It is of interest to understand which of the variables exercises the largest influence of models' predictions. Toward this aim, we can apply the permutation-based variable-importance measure discussed in Chapter 16. Subsequently, we can construct a plot of the obtained mean (over the default 10 permutations) variable-importance measures. Note that we consider only the top-20 variables.

The resulting plot is shown in Figure 21.6. The bar for each explanatory variable starts at the RMSE value of a particular model and ends at the (mean) RMSE calculated for data with permuted values of the variable.

Figure 21.6 indicates that, for the gradient boosting and random forest models, the two explanatory variables with the largest values of the importance measure are `Reactions` or `BallControl`. The importance of other variables varies depending on the model. Interestingly, in the linear-regression model, the highest importance is given to goal-keeping skills.

We may also want to take a look at the partial-dependence (PD) profiles discussed in Chapter 17. Recall that they illustrate how does the expected value of a model's predictions behave as a function of an explanatory variable. To create the profiles, we apply function `model_profile()` from the `DALEX` package (see Section 17.6). We focus on variables `Reactions`, `BallControl`, and `Dribbling` that were important in the random forest model (see Figure 21.6). We also consider `Age`, as it had some effect in the gradient boosting models. Subsequently, we can construct a plot of contrastive PD profiles (see Section 17.3.4) that is shown in Figure 21.7.

Figure 21.7 indicates that the shape of the PD profiles for `Reactions`, `BallControl`, and `Dribbling` is, in general, similar for all the models and implies an increasing predicted player's value for an increasing (at least, after passing some threshold) value of the explanatory variable. However, for `Age`, the shape is different and suggests a decreasing player's value after the age of about 25 years. It is worth noting that the range of expected model's predictions is, in general, the smallest for the random forest model. Also, the three tree-based models tend to stabilize the predictions at the ends of the explanatory-variable ranges.

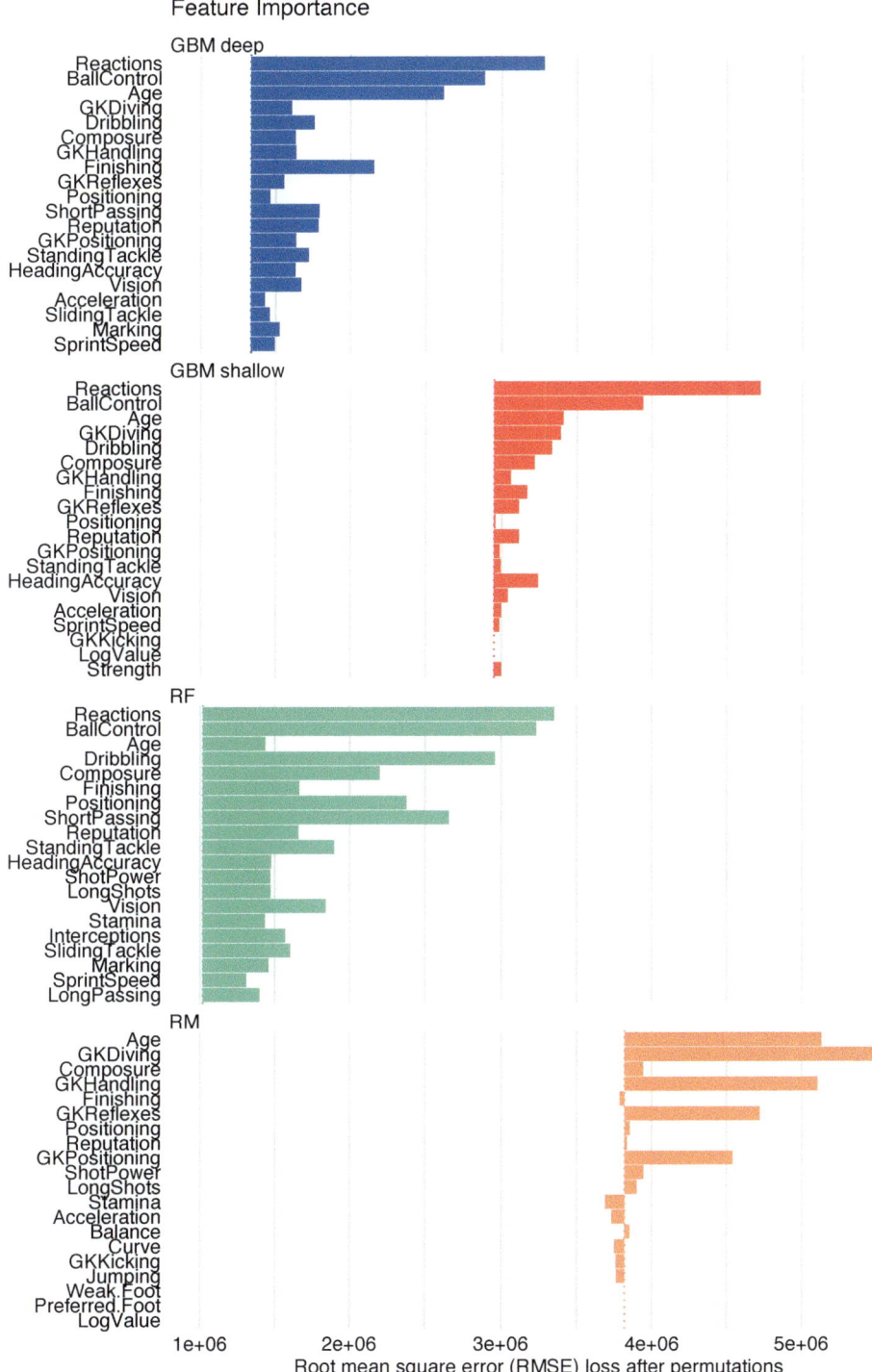

FIGURE 21.6 Mean variable-importance calculated using 10 permutations for the four models for the FIFA 19 data.

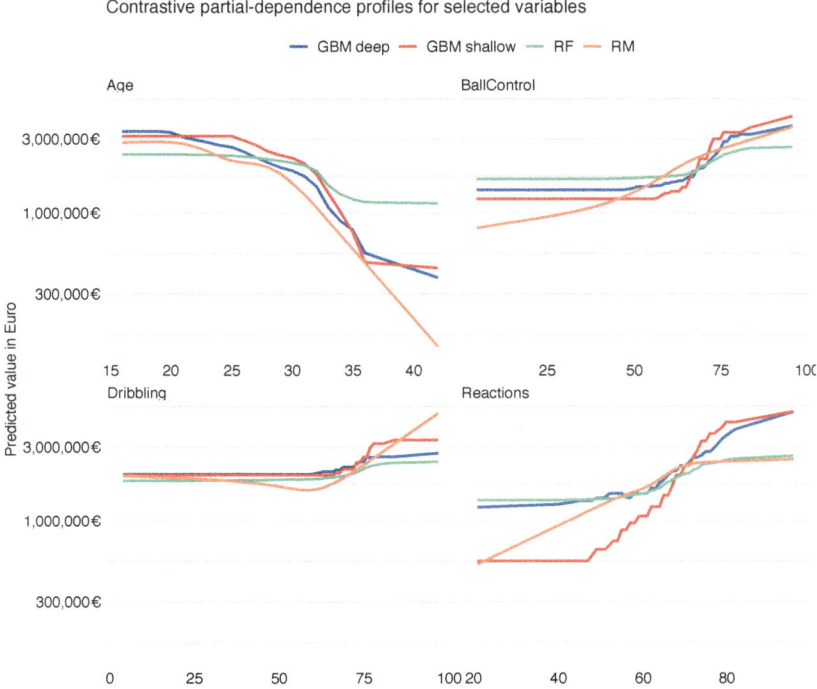

FIGURE 21.7 Contrastive partial-dependence profiles for the four models and selected explanatory variables for the FIFA 19 data.

The most interesting difference between the conclusions drawn from Figure 21.3 and those obtained from Figure 21.7 is observed for variable `Age`. In particular, Figure 21.3 suggests that the relationship between player's age and value is non-monotonic, while Figure 21.7 suggests a non-increasing relationship. How can we explain this difference? A possible explanation is as follows. The youngest players have lower values, not because of their age, but because of their lower skills, which are correlated (as seen from the scatter-plot matrix in Figure 21.4) with young age. The simple data exploration analysis, presented in the upper-left panel of Figure 21.3, cannot separate the effects of age and skills. As a result, the analysis suggests a decrease in player's value for the youngest players. In models, however, the effect of age is estimated while adjusting for the effect of skills. After this adjustment, the effect takes the form of a non-increasing pattern, as shown by the PD profiles for `Age` in Figure 21.7.

This example indicates that *exploration of models may provide more insight than exploration of raw data*. In exploratory data analysis, the effect of variable `Age` was confounded by the effect of skill-related variables. By using a model, the confounding has been removed.

21.6.1   Code snippets for R

In this section, we show R-code snippets for dataset-level exploration for the gradient boosting model. For other models a similar syntax was used.

The `model_parts()` function from the **DALEX** package (see Section 16.6) is used to calculate the permutation-based variable-importance measure. The generic `plot()` function is applied to graphically present the computed values of the measure. The `max_vars` argument is used to limit the number of presented variables up to 20.

```
fifa_mp_gbm_deep <- model_parts(fifa_gbm_exp_deep)
plot(fifa_mp_gbm_deep, max_vars = 20,
     bar_width = 4, show_boxplots = FALSE)
```

The `model_profile()` function from the **DALEX** package (see Section 17.6) is used to calculate PD profiles. The generic `plot()` function is used to graphically present the profiles for selected variables.

```
selected_variables <- c("Reactions", "BallControl", "Dribbling", "Age")
fifa19_pd_deep <- model_profile(fifa_gbm_exp_deep,
                            variables = selected_variables)
plot(fifa19_pd_deep)
```

21.6.2   Code snippets for Python

In this section, we show Python code snippets for dataset-level exploration for the gradient boosting model. For other models a similar syntax was used.

The `model_parts()` method from the **dalex** library (see Section 16.7) is used to calculate the permutation-based variable-importance measure. The `plot()` method is applied to graphically present the computed values of the measure.

```
fifa_mp_gbm = fifa_gbm_exp.model_parts()
fifa_mp_gbm.plot(max_vars = 20)
```

The `model_profile()` method from the **dalex** library (see Section 17.7) is used to calculate PD profiles. The `plot()` method is used to graphically present the computed profiles.

```
fifa_mp_gbm = fifa_gbm_exp.model_profile()

fifa_mp_gbm.plot(variables = ['movement_reactions',
    'skill_ball_control', 'skill_dribbling', 'age'])
```

In order to calculated other types of profiles, just change the `type` argument.

```
fifa_mp_gbm = fifa_gbm_exp.model_profile(type = 'accumulated')

fifa_mp_gbm.plot(variables = ['movement_reactions',
    'skill_ball_control', 'skill_dribbling', 'age'])
```

TABLE 21.3 Characteristics of Robert Lewandowski.

Age	29	Dribbling	85	ShotPower	88	Composure	86
Preferred.Foot	2	Curve	77	Jumping	84	Marking	34
Reputation	4	FKAccuracy	86	Stamina	78	StandingTackle	42
Weak.Foot	4	LongPassing	65	Strength	84	SlidingTackle	19
Skill.Moves	4	BallControl	89	LongShots	84	GKDiving	15
Crossing	62	Acceleration	77	Aggression	80	GKHandling	6
Finishing	91	SprintSpeed	78	Interceptions	39	GKKicking	12
Heading	85	Agility	78	Positioning	91	GKPositioning	8
ShortPassing	83	Reactions	90	Vision	77	GKReflexes	10
Volleys	89	Balance	78	Penalties	88	LogValue	8

21.7   Instance-level explanations

After evaluation of the models at the dataset-level, we may want to focus on particular instances.

21.7.1   Robert Lewandowski

As a first example, we take a look at the value of *Robert Lewandowski*, for an obvious reason. Table 21.3 presents his characteristics, as included in the analyzed dataset. Robert Lewandowski is a striker.

First, we take a look at variable attributions, discussed in Chapter 6. Recall that they decompose model's prediction into parts that can be attributed to different explanatory variables. The attributions can be presented in a breakdown (BD) plot. For brevity, we only consider the random forest model. The resulting BD plot is shown in Figure 21.8.

Figure 21.8 suggests that the explanatory variables with the largest effect are `Composure`, `Volleys`, `LongShots`, and `Stamina`. However, in Chapter 6 it was mentioned that variable attributions may depend on the order of explanatory covariates that are used in calculations. Thus, in Chapter 8 we introduced Shapley values, based on the idea of averaging the attributions over many orderings. Figure 21.9 presents the means of the Shapley values computed by using 25 random orderings for the random forest model.

Figure 21.9 indicates that the five explanatory variables with the largest Shapley values are `BallControl`, `Dribbling`, `Reactions`, `ShortPassing`, and `Positioning`. This makes sense, as Robert Lewandowski is a striker.

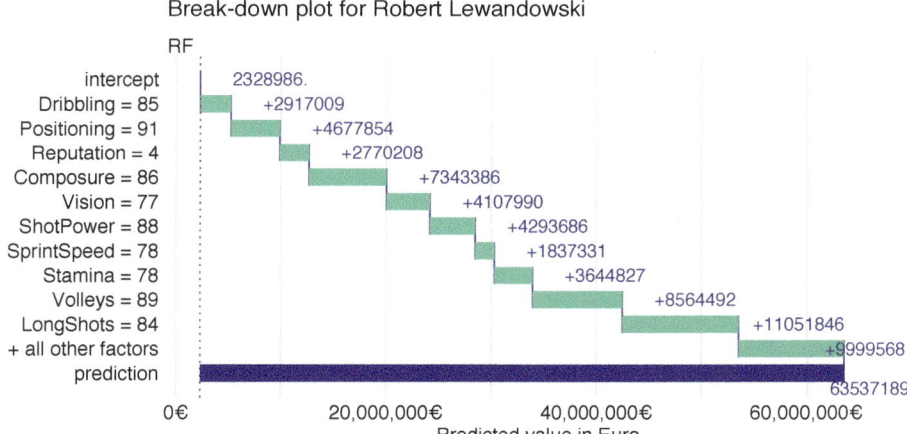

FIGURE 21.8 Break-down plot for Robert Lewandowski for the random forest model.

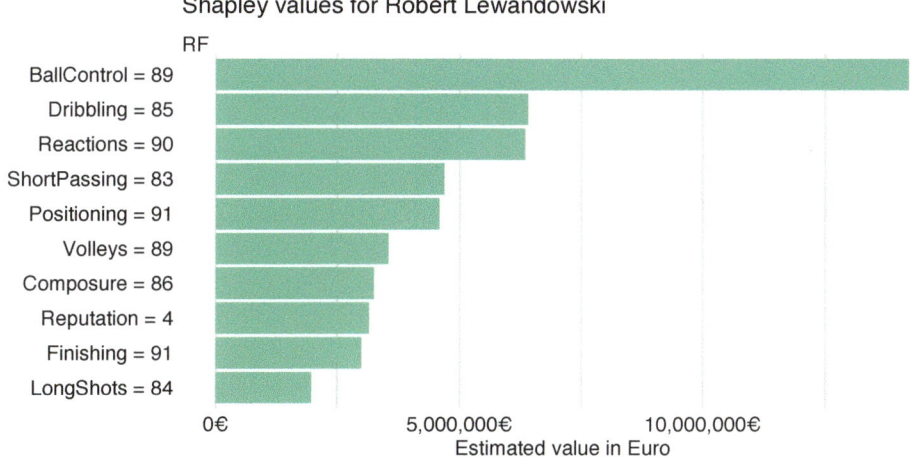

FIGURE 21.9 Shapley values for Robert Lewandowski for the random forest model.

In Chapter 10, we introduced ceteris-paribus (CP) profiles. They capture the effect of a selected explanatory variable in terms of changes in a model's prediction induced by changes in the variable's values. Figure 21.10 presents the profiles for variables Age, Reactions, BallControl, and Dribbling for the random forest model.

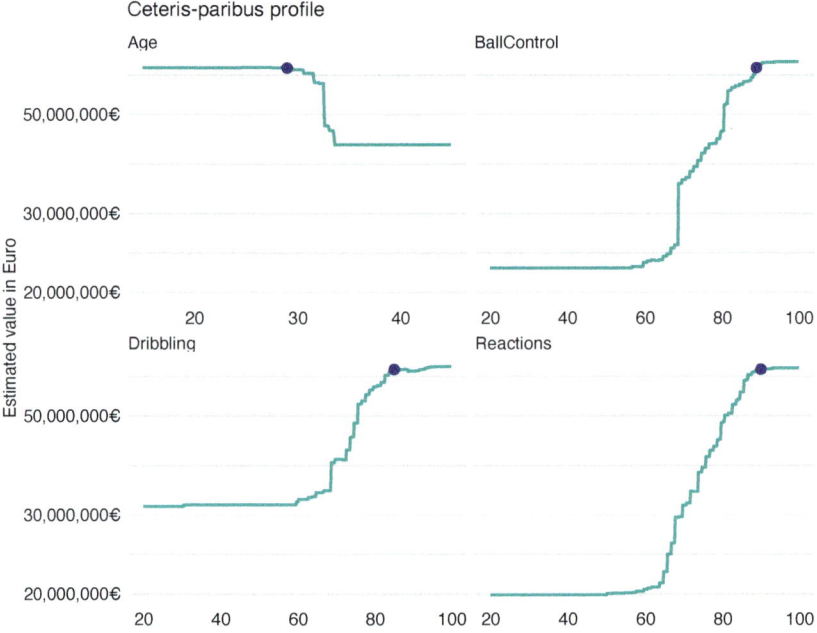

FIGURE 21.10 Ceteris-paribus profiles for Robert Lewandowski for four selected variables and the random forest model.

Figure 21.10 suggests that, among the four variables, BallControl and Reactions lead to the largest changes of predictions for this instance. For all four variables, the profiles flatten at the left- and right-hand-side edges. The predicted value of Robert Lewandowski reaches or is very close to the maximum for all four profiles. It is interesting to note that, for Age, the predicted value is located at the border of the age region at which the profile suggests a sharp drop in player's value.

As it was argued in Chapter 12, it is worthwhile to check how does the model behave for observations similar to the instance of interest. Towards this aim, we may want to compare the distribution of residuals for "neighbors" of Robert Lewandowski. Figure 21.11 presents the histogram of residuals for all data and the 30 neighbors of Robert Lewandowski.

Clearly, the neighbors of Robert Lewandowski include some of the most expensive players. Therefore, as compared to the overall distribution, the distribution of residuals for the neighbors, presented in Figure 21.11, is skewed

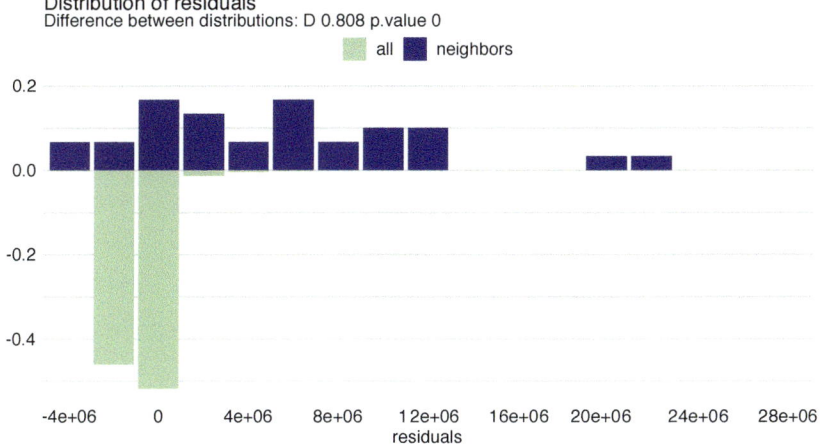

FIGURE 21.11 Distribution of residuals for the random forest model for all players and for 30 neighbors of Robert Lewandowski.

to the right, and its mean is larger than the overall mean. Thus, the model underestimates the actual value of the most expensive players. This was also noted based on the plot in the bottom-left panel of Figure 21.5.

We can also look at the local-stability plot, i.e., the plot that includes CP profiles for the nearest neighbors and the corresponding residuals (see Chapter 12). In Figure 21.12, we present the plot for Age.

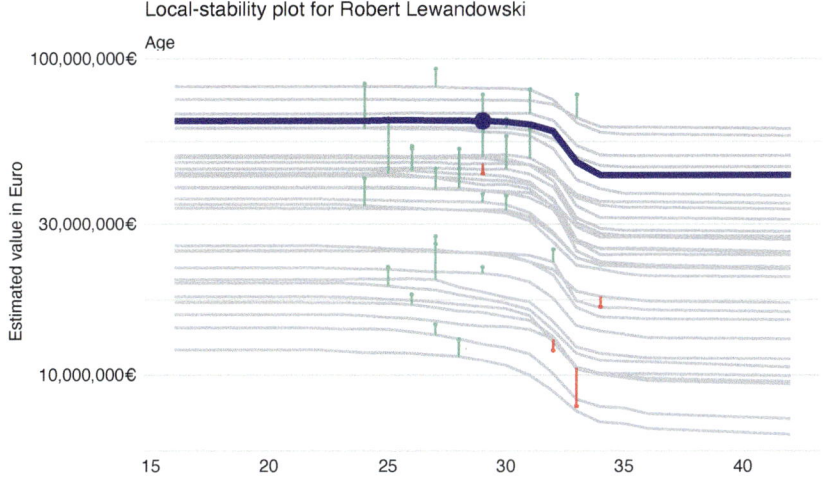

FIGURE 21.12 Local-stability plot for Age for 30 neighbors of Robert Lewandowski and the random forest model.

The CP profiles in Figure 21.12 are almost parallel but span quite a wide range of the predicted player's values. Thus, one could conclude that the predictions for the most expensive players are not very stable. Also, the plot includes more positive residuals (indicated in the plot by green vertical intervals) than negative ones (indicated by red vertical intervals). This confirms the conclusion drawn from Figure 21.11 that the values of the most expensive players are underestimated by the model.

21.7.2   Code snippets for R

In this section, we show R-code snippets for instance-level exploration for the gradient boosting model. For other models, a similar syntax was used.

The `predict_parts()` function from the `DALEX` package (see Chapters 6-8) is used to calculate variable attributions. Note that we apply the `type = "break_down"` argument to prepare BD plots. The generic `plot()` function is used to graphically present the plots.

```
fifa_bd_gbm <- predict_parts(fifa_gbm_exp,
                  new_observation = fifa["R. Lewandowski",],
                  type = "break_down")
plot(fifa_bd_gbm) +
  scale_y_continuous("Predicted value in Euro",
                  labels = dollar_format(suffix = "€", prefix = "")) +
  ggtitle("Break-down plot for Robert Lewandowski","")
```

Shapley values are computed by applying the `type = "shap"` argument.

```
fifa_shap_gbm <- predict_parts(fifa_gbm_exp,
                  new_observation = fifa["R. Lewandowski",],
                  type = "shap")
plot(fifa_shap_gbm, show_boxplots = FALSE) +
  scale_y_continuous("Estimated value in Euro",
                  labels = dollar_format(suffix = "€", prefix = "")) +
  ggtitle("Shapley values for Robert Lewandowski","")
```

The `predict_profile()` function from the `DALEX` package (see Section 10.6) is used to calculate the CP profiles. The generic `plot()` function is applied to graphically present the profiles.

```
selected_variables <- c("Reactions", "BallControl", "Dribbling", "Age")
fifa_cp_gbm <- predict_profile(fifa_gbm_exp,
                  new_observation = fifa["R. Lewandowski",],
                      variables = selected_variables)
plot(fifa_cp_gbm, variables = selected_variables)
```

Finally, the `predict_diagnostics()` function (see Section 12.6) allows calculating local-stability plots. The generic `plot()` function can be used to plot these profiles for selected variables.

```
id_gbm <- predict_diagnostics(fifa_gbm_exp,
                              fifa["R. Lewandowski",],
                              neighbors = 30)
plot(id_gbm) +
  scale_y_continuous("Estimated value in Euro", trans = "log10",
                     labels = dollar_format(suffix = "€", prefix = ""))
```

21.7.3   Code snippets for Python

In this section, we show Python-code snippets for instance-level exploration for the gradient boosting model. For other models, a similar syntax was used.

First, we need to select instance of interest. In this example we will use *Cristiano Ronaldo*.

```
cr7 = X.loc['Cristiano Ronaldo',]
```

The `predict_parts()` method from the **dalex** library (see Sections 6.7 and 8.6) can be used to calculate calculate variable attributions. The `plot()` method with `max_vars` argument is applied to graphically present the corresponding BD plot for up to 20 variables.

```
fifa_pp_gbm = fifa_gbm_exp.predict_parts(cr7, type='break_down')
fifa_pp_gbm.plot(max_vars = 20)
```

To calculate Shapley values, the `predict_parts()` method should be applied with the `type='shap'` argument (see Section 8.6).

The `predict_profile()` method from the **dalex** library (see Section 10.7) allows calculation of the CP profiles. The `plot()` method with the `variables` argument plots the profiles for selected variables.

```
fifa_mp_gbm = fifa_gbm_exp.predict_profile(cr7)

fifa_mp_gbm.plot(variables =  ['movement_reactions',
    'skill_ball_control', 'skill_dribbling', 'age'])
```

21.7.4   CR7

As a second example, we present explanations for the random forest-model's prediction for *Cristiano Ronaldo* (CR7). Table 21.4 presents his characteristics, as included in the analyzed dataset. Note that Cristiano Ronaldo, as Robert Lewandowski, is also a striker. It might be thus of interest to compare the characteristics contributing to the model's predictions for the two players.

The BD plot for Cristiano Ronaldo is presented in Figure 21.13. It suggests that the explanatory variables with the largest effect are `ShotPower`, `LongShots`, `Volleys`, and `Vision`.

TABLE 21.4 Characteristics of Cristiano Ronaldo.

Age	33	Dribbling	88	ShotPower	95	Composure	95
Preferred.Foot	2	Curve	81	Jumping	95	Marking	28
Reputation	5	FKAccuracy	76	Stamina	88	StandingTackle	31
Weak.Foot	4	LongPassing	77	Strength	79	SlidingTackle	23
Skill.Moves	5	BallControl	94	LongShots	93	GKDiving	7
Crossing	84	Acceleration	89	Aggression	63	GKHandling	11
Finishing	94	SprintSpeed	91	Interceptions	29	GKKicking	15
Heading	89	Agility	87	Positioning	95	GKPositioning	14
ShortPassing	81	Reactions	96	Vision	82	GKReflexes	11
Volleys	87	Balance	70	Penalties	85	LogValue	8

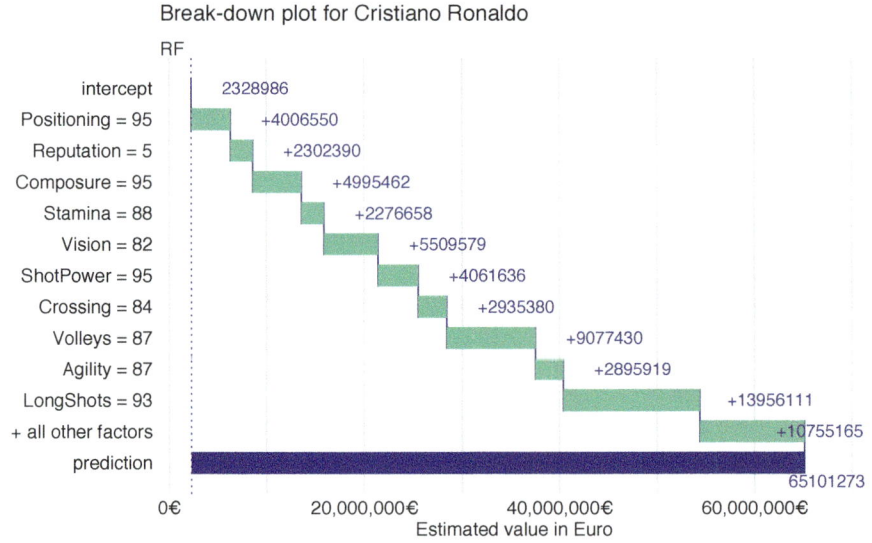

FIGURE 21.13 Break-down plot for Cristiano Ronaldo for the random forest model.

Figure 21.14 presents Shapley values for Cristiano Ronaldo. It indicates that the four explanatory variables with the largest values are `Reactions`, `Dribbling`, `BallControl`, and `ShortPassing`. These are the same variables as for Robert Lewandowski, though in a different order. Interestingly, the plot for Cristiano Ronaldo includes variable `Age`, for which Shapley value is negative. It suggests that CR7's age has got a negative effect on the model's prediction.

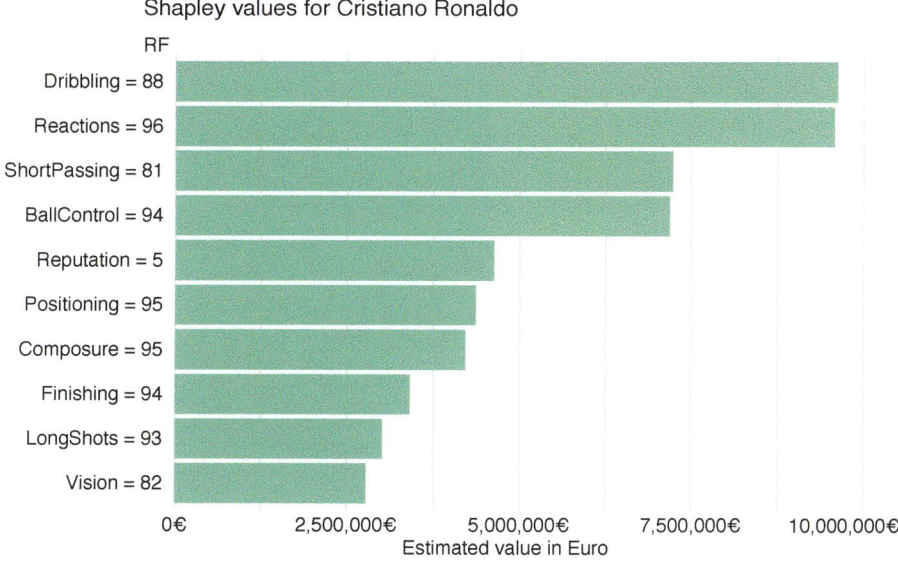

FIGURE 21.14 Shapley values for Cristiano Ronaldo for the random forest model.

Finally, Figure 21.15 presents CP profiles for `Age`, `Reactions`, `Dribbling`, and `BallControl`.

The profiles are similar to those presented in Figure 21.10 for Robert Lewandowski. An interesting difference is that, for `Age`, the predicted value for Cristiano Ronaldo is located within the region of age, linked with a sharp drop in player's value. This is in accordance with the observation, made based on Figure 21.14, that CR7's age has got a negative effect on the model's prediction.

21.7.5 Wojciech Szczęsny

One might be interested in the characteristics influencing the random forest model's predictions for players other than strikers. To address the question, we present explanations for *Wojciech Szczęsny*, a goalkeeper. Table 21.5 presents his characteristics, as included in the analyzed dataset.

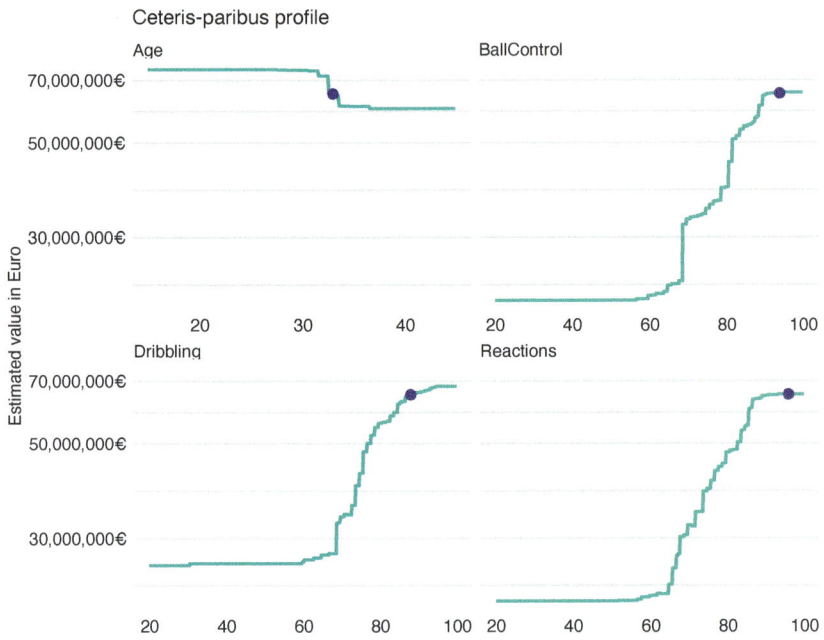

FIGURE 21.15 Ceteris-paribus profiles for Cristiano Ronaldo for four selected variables and the random forest model.

TABLE 21.5 Characteristics of Wojciech Szczęsny.

Age	28	Dribbling	11	ShotPower	15	Composure	65
Preferred.Foot	2	Curve	16	Jumping	71	Marking	20
Reputation	3	FKAccuracy	14	Stamina	45	StandingTackle	13
Weak.Foot	3	LongPassing	36	Strength	65	SlidingTackle	12
Skill.Moves	1	BallControl	22	LongShots	14	GKDiving	85
Crossing	12	Acceleration	51	Aggression	40	GKHandling	81
Finishing	12	SprintSpeed	47	Interceptions	15	GKKicking	71
Heading	16	Agility	55	Positioning	14	GKPositioning	85
ShortPassing	32	Reactions	82	Vision	48	GKReflexes	87
Volleys	14	Balance	51	Penalties	18	LogValue	8

Figure 21.16 shows the BD plot. We can see that the most important contributions come from the explanatory variables related to goalkeeping skills like GKPositioning, GKHandling, and GKReflexes. Interestingly, field-player skills like BallControl or Dribbling have a negative effect.

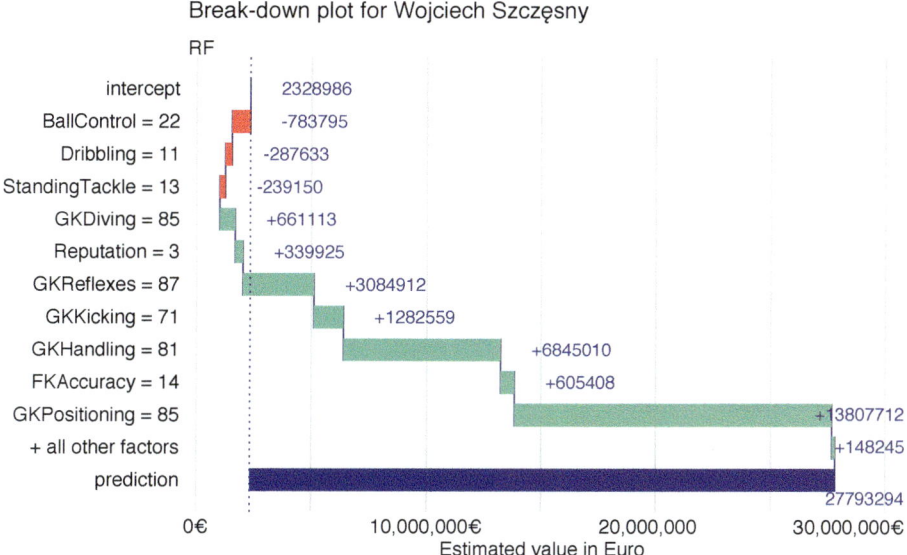

FIGURE 21.16 Break-down plot for Wojciech Szczęsny for the random forest model.

Figure 21.17 presents Shapley values (over 25 random orderings of explanatory variables). The plot confirms that the most important contributions to the prediction for Wojciech Szczęsny are due to goalkeeping skills like GKDiving, GKPositioning, GKReflexes, and GKHandling. Interestingly, Reactions is also important, as it was the case for Robert Lewandowski (see Figure 21.9) and Cristiano Ronaldo (see Figure 21.14).

21.7.6 Lionel Messi

This instance might be THE choice for some of the readers. However, we have decided to leave explanation of the models' predictions in this case as an exercise to the interested readers.

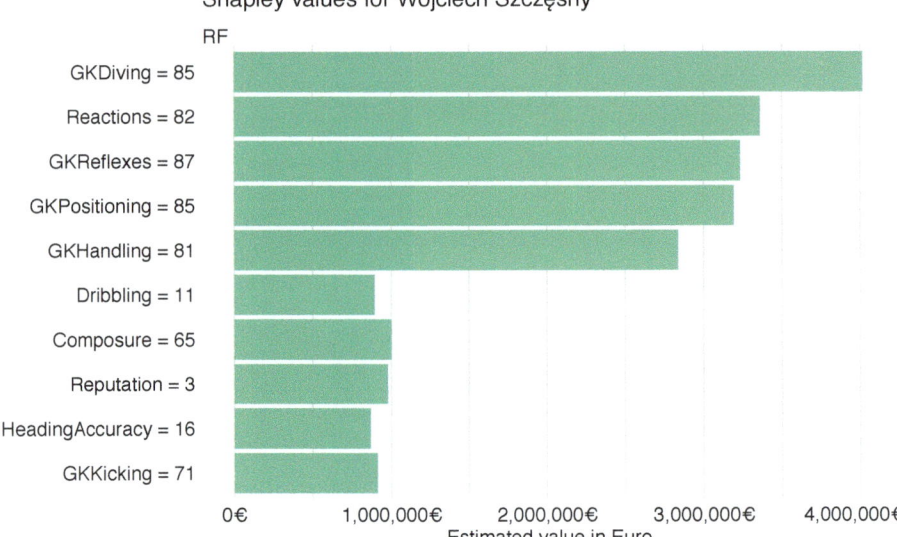

FIGURE 21.17 Shapley values for Wojciech Szczęsny for the random forest model.

22

Reproducibility

All examples presented in this book are reproducible. Parts of the source codes are available in the book. Over time, some of the functionality of the described packages may change. The online version of the book is updated. Fully reproducible code is available at https://pbiecek.github.io/ema/.

Possible differences may be caused by using different versions of the installed packages. Results in this version of the book are obtained with the following versions of the packages.

22.1 Package versions for R

The current versions of packages in R can be checked with `sessionInfo()`.

```
sessionInfo()
```

```
## R version 4.0.2 (2020-06-22)
## Platform: x86_64-apple-darwin17.0 (64-bit)
## Running under: macOS Catalina 10.15.7
##
## locale:
## [1] en_US.UTF-8/en_US.UTF-8/en_US.UTF-8/C/en_US.UTF-8/en_US.UTF-8
##
## attached base packages:
## [1] grid      stats     graphics  grDevices utils     datasets
## [7] methods   base
##
## other attached packages:
##  [1] ranger_0.12.1         GGally_2.0.0          tidyr_1.1.2
##  [4] scales_1.1.1          tree_1.0-40           gridExtra_2.3
##  [7] dplyr_1.0.2           caret_6.0-86          ingredients_2.0
## [10] gower_0.2.2           glmnet_4.0-2          Matrix_1.2-18
## [13] iml_0.10.1            localModel_0.5        lime_0.5.1
## [16] DALEXtra_2.0          iBreakDown_1.3.1.9000 e1071_1.7-4
## [19] gbm_2.1.8             randomForest_4.6-14   rms_6.0-1
## [22] SparseM_1.78          Hmisc_4.4-1           Formula_1.2-4
## [25] survival_3.2-7        lattice_0.20-41       forcats_0.5.0
```

```
## [28] patchwork_1.0.1       ggmosaic_0.3.0        kableExtra_1.2.1
## [31] knitr_1.30            DALEX_2.0.1           ggplot2_3.3.2
##
## loaded via a~namespace (and not attached):
##   [1] backports_1.1.10     workflows_0.2.1      plyr_1.8.6
##   [4] lazyeval_0.2.2       splines_4.0.2        listenv_0.8.0
##   [7] TH.data_1.0-10       digest_0.6.27        foreach_1.5.1
##  [10] htmltools_0.5.0      parsnip_0.1.3        productplots_0.1.1
##  [13] fansi_0.4.1          memoise_1.1.0        magrittr_1.5
##  [16] checkmate_2.0.0      cluster_2.1.0        Metrics_0.1.4
##  [19] recipes_0.1.14       globals_0.13.1       matrixStats_0.57.0
##  [22] sandwich_3.0-0       dials_0.0.9          jpeg_0.1-8.1
##  [25] colorspace_2.0-0     blob_1.2.1           rvest_0.3.6
##  [28] xfun_0.18            RCurl_1.98-1.2       crayon_1.3.4
##  [31] jsonlite_1.7.1       libcoin_1.0-6        flock_0.7
##  [34] zoo_1.8-8            iterators_1.0.13     glue_1.4.2
##  [37] gtable_0.3.0         ipred_0.9-9          webshot_0.5.2
##  [40] MatrixModels_0.4-1   shape_1.4.5          mvtnorm_1.1-1
##  [43] DBI_1.1.0            Rcpp_1.0.5           archivist_2.3.4
##  [46] viridisLite_0.3.0    xtable_1.8-4         htmlTable_2.1.0
##  [49] reticulate_1.16      bit_4.0.4            GPfit_1.0-8
##  [52] foreign_0.8-80       stats4_4.0.2         prediction_0.3.14
##  [55] lava_1.6.8           prodlim_2019.11.13   htmlwidgets_1.5.2
##  [58] httr_1.4.2           RColorBrewer_1.1-2   ellipsis_0.3.1
##  [61] reshape_0.8.8        pkgconfig_2.0.3      farver_2.0.3
##  [64] nnet_7.3-14          reshape2_1.4.4       DiceDesign_1.8-1
##  [67] tidyselect_1.1.0     labeling_0.4.2       rlang_0.4.8
##  [70] later_1.1.0.1        munsell_0.5.0        tools_4.0.2
##  [73] cli_2.1.0            RSQLite_2.2.1        generics_0.1.0
##  [76] evaluate_0.14        stringr_1.4.0        fastmap_1.0.1
##  [79] yaml_2.2.1           bit64_4.0.5          ModelMetrics_1.2.2.2
##  [82] purrr_0.3.4          future_1.19.1        nlme_3.1-149
##  [85] mime_0.9             quantreg_5.73        xml2_1.3.2
##  [88] compiler_4.0.2       shinythemes_1.1.2    rstudioapi_0.13
##  [91] plotly_4.9.2.1       png_0.1-7            tibble_3.0.4
##  [94] lhs_1.1.1            stringi_1.5.3        highr_0.8
##  [97] vctrs_0.3.4          pillar_1.4.6         lifecycle_0.2.0
## [100] bitops_1.0-6         data.table_1.13.2    conquer_1.0.2
## [103] httpuv_1.5.4         R6_2.5.0             latticeExtra_0.6-29
## [106] bookdown_0.21        promises_1.1.1       codetools_0.2-16
## [109] polspline_1.1.19     MASS_7.3-53          assertthat_0.2.1
## [112] withr_2.3.0          multcomp_1.4-14      mgcv_1.8-33
## [115] parallel_4.0.2       rpart_4.1-15         timeDate_3043.102
## [118] class_7.3-17         rmarkdown_2.4        inum_1.0-1
## [121] pROC_1.16.2          partykit_1.2-10      shiny_1.5.0
## [124] lubridate_1.7.9      base64enc_0.1-3
```

22.2 Package versions for Python

The current versions of packages in Python can be checked with `pip`.

```
pip freeze
```

```
absl-py==0.6.1              appnope==0.1.0
astor==0.7.1               atomicwrites==1.2.1
attrs==18.2.0             backcall==0.1.0
bleach==3.1.1            cycler==0.10.0
dalex==0.3.0              dbexplorer==1.21
decorator==4.4.2          defusedxml==0.6.0
entrypoints==0.3          future==0.17.1
gast==0.2.0               grpcio==1.16.1
h5py==2.8.0               imageio==2.9.0
importlib-metadata==1.5.0 innvestigate==1.0.4
ipykernel==5.1.4          ipython==7.13.0
ipython-genutils==0.2.0   jedi==0.16.0
Jinja2==2.11.1            joblib==0.14.1
json5==0.9.2              jsonschema==3.2.0
jupyter-client==6.0.0     jupyter-core==4.6.3
jupyterlab==2.0.1         jupyterlab-server==1.0.7
Keras==2.2.2              Keras-Applications==1.0.6
Keras-Preprocessing==1.0.5 kiwisolver==1.0.1
lightgbm==2.3.1           lime==0.2.0.1
Markdown==3.0.1           MarkupSafe==1.1.1
matplotlib==2.2.2         mistune==0.8.4
more-itertools==4.3.0     nbconvert==5.6.1
nbformat==5.0.4           networkx==2.4
notebook==6.0.3           numpy==1.19.0
pandas==1.1.1             pandocfilters==1.4.2
parso==0.6.2              patsy==0.5.1
pd==0.0.1                 pexpect==4.8.0
pickleshare==0.7.5        Pillow==5.3.0
plotly==4.9.0             pluggy==0.8.0
prometheus-client==0.7.1  prompt-toolkit==3.0.3
protobuf==3.6.1           psycopg2==2.7.4
ptyprocess==0.6.0         py==1.7.0
Pygments==2.5.2           PyMySQL==0.8.0
pyodbc==4.0.23            pyparsing==2.2.0
pyrsistent==0.15.7        pytest==4.0.0
python-dateutil==2.7.3    pytz==2018.4
PyWavelets==1.1.1         PyYAML==3.13
pyzmq==19.0.0             retrying==1.3.3
scikit-image==0.17.2      scikit-learn==0.22.2.post1
scipy==1.1.0              Send2Trash==1.5.0
simplejson==3.14.0        six==1.11.0
```

```
sklearn==0.0                    sphinx-rtd-theme==0.4.0
statsmodels==0.11.1             tabulate==0.8.7
tensorboard==1.12.0             tensorflow==1.12.0
termcolor==1.1.0                terminado==0.8.3
testpath==0.4.4                 tifffile==2020.7.17
torch==1.0.1.post2              torchvision==0.2.1
tornado==6.0.4                  tqdm==4.43.0
traitlets==4.3.3                typing==3.6.4
virtualenv==16.2.0              wcwidth==0.1.8
webencodings==0.5.1             Werkzeug==0.14.1
xgboost==0.72                   zipp==3.1.0
```

Bibliography

Alber, M., Lapuschkin, S., Seegerer, P., Hägele, M., Schütt, K. T., Montavon, G., Samek, W., Müller, K.-R., Dähne, S., and Kindermans, P.-J. (2019). iNNvestigate Neural Networks! *Journal of Machine Learning Research*, 20(93):1–8.

Allaire, J. and Chollet, F. (2019). *keras: R Interface to Keras*. R package version 2.2.4.1.

Allison, P. (2014). Measures of fit for logistic regression. In *Proceedings of the SAS Global Forum 2014 Conference*, Cary, NC. SAS Institute Inc.

Alvarez-Melis, D. and Jaakkola, T. S. (2018). On the Robustness of Interpretability Methods. *ICML Workshop on Human Interpretability in Machine Learning (WHI 2018)*.

Apley, D. (2018). *ALEPlot: Accumulated Local Effects (ALE) Plots and Partial Dependence (PD) Plots*. R package version 1.1.

Apley, D. W. and Zhu, J. (2020). Visualizing the effects of predictor variables in black box supervised learning models. *Journal of the Royal Statistical Society Series B*, 82(4):1059–1086.

Azure (2019). *Microsoft Cognitive Services*.

Bach, S., Binder, A., Montavon, G., Klauschen, F., Müller, K.-R., and Samek, W. (2015). On pixel-wise explanations for non-linear classifier decisions by layer-Wise relevance propagation. *Plos One*, 10(7):e0130140.

Berrar, D. (2019). Performance measures for binary classification. In *Encyclopedia of Bioinformatics and Computational Biology Volume 1*, pages 546–560. Elsevier.

Biecek, P. (2018). DALEX: Explainers for complex predictive models in R. *Journal of Machine Learning Research*, 19(84):1–5.

Biecek, P. (2019). Model Development Process. *CoRR*, abs/1907.04461.

Biecek, P., Baniecki, H., Izdebski, A., and Pekala, K. (2019). *ingredients: Effects and Importances of Model Ingredients*. R package version 1.3.3.

Biecek, P. and Kosinski, M. (2017). archivist: An R package for managing, recording and restoring data analysis results. *Journal of Statistical Software*, 82(11):1–28.

Binder, A., Montavon, G., Bach, S., Müller, K., and Samek, W. (2016). Layer-wise relevance propagation for neural networks with local renormalization layers. In *Artificial Neural Networks and Machine Learning - 25th International Conference on Artificial Neural Networks, ICANN 2016, Proceedings*, volume 9887 LNCS of *Lecture Notes in Computer Science*, pages 63–71. Springer Verlag. 25th International Conference on Artificial Neural Networks and Machine Learning, ICANN 2016.

Bischl, B., Lang, M., Kotthoff, L., Schiffner, J., Richter, J., Studerus, E., Casalicchio, G., and Jones, Z. M. (2016). mlr: Machine Learning in R. *Journal of Machine Learning Research*, 17(170):1–5.

Boehm, B. (1988). A spiral model of software development and enhancement. *IEEE Computer, IEEE*, 21(5):61–72.

Breiman, L. (2001a). Random forests. *Machine Learning*, 45:5–32.

Breiman, L. (2001b). Statistical modeling: The two cultures. *Statistical Science*, 16(3):199–231.

Breiman, L., Cutler, A., Liaw, A., and Wiener, M. (2018). *randomForest: Breiman and Cutler's Random Forests for Classification and Regression*. R package version 4.6-14.

Breiman, L., Friedman, J. H., Olshen, R. A., and Stone, C. J. (1984). *Classification and Regression Trees*. Wadsworth and Brooks, Monterey, CA.

Brentnall, A. and Cuzick, J. (2018). Use of the concordance index for predictors of censored survival data. *Statistical Methods in Medical Research*, 27:2359–2373.

Casey, B., Farhangi, A., and Vogl, R. (2019). Rethinking explainable machines: The GDPR's Right to Explanation debate and the rise of algorithmic audits in enterprise. *Berkeley Technology Law Journal*, 34:143–188.

Chapman, P., Clinton, J., Kerber, R., Khabaza, T., Reinartz, T., Shearer, C., and Wirth, R. (1999). *The CRISP-DM 1.0 Step-by-step data mining guide*.

Chen, T. and Guestrin, C. (2016). XGBoost: A scalable tree boosting system. In *Proceedings of the 22nd ACM SIGKDD International Conference on Knowledge Discovery and Data Mining*, KDD '16, pages 785–794. ACM.

Cortes, C. and Vapnik, V. (1995). Support-vector networks. *Machine Learning*, 20:273–297.

Dastin, J. (2018). Amazon scraps secret AI recruiting tool that showed bias against women. *Reuters*. https://reut.rs/2UhvfZP.

Deng, J., Dong, W., Socher, R., Li, L., Kai Li, and Li Fei-Fei (2009). ImageNet: A large-scale hierarchical image database. In *2009 IEEE Conference on Computer Vision and Pattern Recognition*, pages 248–255, Los Alamitos, CA, USA. IEEE Computer Society.

Diaz, M., Johnson, I., Lazar, A., Piper, A. M., and Gergle, D. (2018). Addressing age-related bias in sentiment analysis. In *Proceedings of the 2018 CHI Conference on Human Factors in Computing Systems*, Chi '18, pages 412:1–412:14. ACM.

Dobson, A. (2002). *Introduction to Generalized Linear Models (2nd Ed.)*. Chapman and Hall/CRC, Boca Raton, FL.

Donizy, P., Biecek, P., Halon, A., and Matkowski, R. (2016). BILLCD8 – a multivariable survival model as a simple and clinically useful prognostic tool to identify high-risk cutaneous melanoma patients. *Anticancer Research*, 36:4739–4748.

Dorogush, A. V., Ershov, V., and Gulin, A. (2018). CatBoost: gradient boosting with categorical features support. *CoRR*, abs/1810.11363.

Duffy, C. (2019). Apple co-founder Steve Wozniak says Apple Card discriminated against his wife. *CNN Business*. https://cnn.it/36i6kLq.

Efron, B. and Hastie, T. (2016). *Computer Age Statistical Inference: Algorithms, Evidence, and Data Science (1st Ed.)*. Cambridge University Press, New York, NY.

Ehrlinger, J. (2016). *ggRandomForests: Exploring Random Forest Survival*. R package version 2.0.1.

Faraway, J. (2005). *Linear Models With R (1st Ed.)*. Chapman and Hall/CRC, Boca Raton, Florida.

Fisher, A., Rudin, C., and Dominici, F. (2019). All models are wrong, but many are useful: Learning a variable's importance by studying an entire class of prediction models simultaneously. *Journal of Machine Learning Research*, 20(177):1–81.

Foster, D. (2017). *xgboostExplainer: An R package that makes xgboost models fully interpretable*. R package version 0.1.

Friedman, J. H. (2000). Greedy function approximation: A gradient boosting machine. *Annals of Statistics*, 29:1189–1232.

Galecki, A. and Burzykowski, T. (2013). *Linear Mixed-Effects Models Using R: A Step-by-Step Approach*. Springer-Verlag New York, New York, NY.

GDPR (2018). *The EU General Data Protection Regulation (GDPR) is the most important change in data privacy regulation in 20 years*.

Goldstein, A., Kapelner, A., Bleich, J., and Pitkin, E. (2015). Peeking inside the black box: Visualizing statistical learning with plots of individual conditional expectation. *Journal of Computational and Graphical Statistics*, 24(1):44–65.

Goodman, B. and Flaxman, S. (2017). European Union regulations on algorithmic decision-making and a "right to explanation". *AI Magazine*, 38(3):50–57.

Gosiewska, A. and Biecek, P. (2018). *auditor: Model Audit - Verification, Validation, and Error Analysis*. R package version 0.2.1.

Gosiewska, A. and Biecek, P. (2019). *iBreakDown: Uncertainty of Model Explanations for Non-additive Predictive Models*. R package version 1.3.3.

Greenwell, B. (2020). *fastshap: Fast Approximate Shapley Values*. R package version 0.0.5.

Greenwell, B. M. (2017). pdp: An R package for constructing partial dependence plots. *The R Journal*, 9(1):421–436.

Grolemund, G. and Wickham, H. (2017). *R for Data Science: Import, Tidy, Transform, Visualize, and Model Data*.

Gulli, A. and Pal, S. (2017). *Deep Learning with Keras*. Packt Publishing Ltd, Birmingham, UK.

Hall, P., Gill, N., and Schmidt, N. (2019). Proposed guidelines for the responsible use of explainable machine learning. *arXiv*, 1906.03533.

Harrell, F. J. (2015). *Regression Modeling Strategies (2nd Ed.)*. Springer, Cham, Switzerland.

Harrell, F. J., Lee, K., and Mark, D. (1996). Multivariable prognostic models: issues in developing models, evaluating assumptions and adequacy, and measuring and reducing errors. *Statistics in Medicine*, 15:361–387.

Harrell Jr, F. E. (2018). *rms: Regression Modeling Strategies*. R package version 5.1-2.

Hastie, T., Tibshirani, R., and Friedman, J. (2009). *The Elements of Statistical Learning. Data Mining, Inference, and Prediction (2nd Ed.)*. Springer, New York, NY.

Hochreiter, S. and Schmidhuber, J. (1997). Long short-term memory. *Neural Computation*, 9(8):1735–1780.

Hoover, B., Strobelt, H., and Gehrmann, S. (2020). exBERT: A visual analysis tool to explore learned representations in Transformer models. In *Proceedings of the 58th Annual Meeting of the Association for Computational Linguistics: System Demonstrations*, pages 187–196, Online. Association for Computational Linguistics.

Hothorn, T., Hornik, K., and Zeileis, A. (2006). Unbiased recursive partitioning: A conditional inference framework. *Journal of Computational and Graphical Statistics*, 15(3):651–674.

Jacobson, I., Booch, G., and Rumbaugh, J. (1999). *The Unified Software Development Process*. Addison-Wesley, Boston, MA.

James, G., Witten, D., Hastie, T., and Tibshirani, R. (2014). *An Introduction to Statistical Learning: With Applications in R*. Springer, New York, NY.

Jiangchun, L. (2018). *Python partial dependence plot toolbox*. Python package version 0.2.0.

Karbowiak, E. and Biecek, P. (2019). *EIX: Explain Interactions in Gradient Boosting Models*. R package version 1.0.

Kruchten, P. (1998). *The Rational Unified Process*. Addison-Wesley.

Kuhn, M. (2008). Building predictive models in R using the caret package. *Journal of Statistical Software*, 28(5):1–26.

Kuhn, M. and Johnson, K. (2013). *Applied Predictive Modeling*. Springer, New York, NY.

Kuhn, M. and Vaughan, D. (2019). *parsnip: A Common API to Modeling and Analysis Functions*. R package version 0.0.2.

Kutner, M., Nachtsheim, C., Neter, J., and Li, W. (2005). *Applied Linear Statistical Models*. McGraw-Hill/Irwin, New York.

Landram, F., Abdullat, A., and Shah, V. (2005). The coefficient of prediction for model specification. *Southwestern Economic Review*, 32:149–156.

Larson, J., Mattu, S., Kirchner, L., and Angwin, J. (2016). How we analyzed the COMPAS recidivism algorithm. *ProPublica*.

Lazer, D., Kennedy, R., King, G., and Vespignani, A. (2014). The parable of Google Flu: traps in big data analysis. *Science*, 343(6176):1203–1205.

LeDell, E., Gill, N., Aiello, S., Fu, A., Candel, A., Click, C., Kraljevic, T., Nykodym, T., Aboyoun, P., Kurka, M., and Malohlava, M. (2019). *h2o: R Interface for H2O*. R package version 3.22.1.1.

Liaw, A. and Wiener, M. (2002). Classification and regression by randomForest. *R News*, 2(3):18–22.

Little, R. and Rubin, D. B. (2002). *Statistical Analysis With Missing Data (2nd Ed.)*. Wiley, Hoboken, NJ.

Lundberg, S. (2019). *SHAP (SHapley Additive exPlanations)*. Python package.

Lundberg, S. M., Erion, G. G., and Lee, S. (2018). Consistent Individualized Feature Attribution for Tree Ensembles. *CoRR*, abs/1802.03888.

Lundberg, S. M. and Lee, S.-I. (2017). A unified approach to interpreting model predictions. In Guyon, I., Luxburg, U. V., Bengio, S., Wallach, H., Fergus, R., Vishwanathan, S., and Garnett, R., editors, *Advances in Neural Information Processing Systems 30*, pages 4765–4774. Curran Associates, Montreal.

Maksymiuk, S., Gosiewska, A., and Biecek, P. (2019). *shapper: Wrapper of Python library shap*. R package version 0.1.2.

Max, K. and Wickham, H. (2018). *tidymodels: Easily Install and Load the 'Tidymodels' Packages*. R package version 0.0.2.

Meyer, D., Dimitriadou, E., Hornik, K., Weingessel, A., and Leisch, F. (2019). *e1071: manual Functions of the Department of Statistics, Probability Theory Group (Formerly: E1071), TU Wien*. R package version 1.7-2.

Molenberghs, G. and Kenward, M. G. (2007). *Missing Data in Clinical Studies*. Wiley, Chichester, England.

Molnar, C. (2019). *Interpretable Machine Learning*. https://christophm.githu b.io/interpretable-ml-book/.

Molnar, C., Bischl, B., and Casalicchio, G. (2018). iml: An R package for Interpretable Machine Learning. *Journal of Open Source Software*, 3(26):786.

Nagelkerke, N. (1991). A note on a general definition of the coefficient of determination. *Biometrika*, 78:691–692.

Nolan, D. and Lang, D. T. (2015). *Data Science in R: A Case Studies Approach to Computational Reasoning and Problem Solving*. Chapman and Hall/CRC, New York, NY.

O'Connell, M., Hurley, C., and Domijan, K. (2017). Conditional visualization for statistical models: an introduction to the condvis package in R. *Journal of Statistical Software, Articles*, 81(5):1–20.

Olhede, S. and Wolfe, P. (2018). The AI spring of 2018. *Significance*, 15(3):6–7.

O'Neil, C. (2016). *Weapons of Math Destruction: How Big Data Increases Inequality and Threatens Democracy*. Crown Publishing Group, New York, NY.

Paluszynska, A. and Biecek, P. (2017). *randomForestExplainer: A set of tools to understand what is happening inside a Random Forest*. R package version 0.9.

Pedersen, T. L. and Benesty, M. (2019). *lime: Local Interpretable Model-Agnostic Explanations*. R package version 0.5.0.

Pedregosa, F., Varoquaux, G., Gramfort, A., Michel, V., Thirion, B., Grisel, O., Blondel, M., Prettenhofer, P., Weiss, R., Dubourg, V., Vanderplas, J., Passos, A., Cournapeau, D., Brucher, M., Perrot, M., and Duchesnay, E. (2011).

Scikit-learn: Machine learning in Python. *Journal of Machine Learning Research*, 12:2825–2830.

Plotly Technologies Inc. (2015). *Collaborative data science.* Montreal, QC.

R Core Team (2018). *R: A Language and Environment for Statistical Computing.* R Foundation for Statistical Computing, Vienna, Austria.

Ribeiro, M. T., Singh, S., and Guestrin, C. (2016). "Why should I trust you?": explaining the predictions of any classifier. In *Proceedings of the 22nd ACM SIGKDD International Conference on Knowledge Discovery and Data Mining, KDD San Francisco, CA*, pages 1135–1144, New York, NY. Association for Computing Machinery.

Ridgeway, G. (2017). *gbm: Generalized Boosted Regression Models.* R package version 2.1.3.

Robnik-Šikonja, M. (2018). *ExplainPrediction: Explanation of Predictions for Classification and Regression Models.* R package version 1.3.0.

Robnik-Šikonja, M. and Kononenko, I. (2008). Explaining classifications for individual instances. *IEEE Transactions on Knowledge and Data Engineering*, 20(5):589–600.

Ross, C. and Swetliz, I. (2018). IBM's Watson supercomputer recommended 'unsafe and incorrect' cancer treatments, internal documents show. *Statnews.* https://bit.ly/38mVxSW.

Rufibach, K. (2010). Use of Brier score to assess binary predictions. *Journal of Clinical Epidemiology*, 63:938–939.

Ruiz, J. (2018). Machine learning and the right to explanation in GDPR. *Open Rights Group.* https://bit.ly/3lgl7ww.

Salzberg, S. (2014). Why Google Flu is a failure. *Forbes.* https://bit.ly/3nc21bc.

Samek, W., Wiegand, T., and Müller, K.-R. (2018). *Explainable Artificial Intelligence: Understanding, Visualizing and Interpreting Deep Learning Models.*

Schafer, J. (1997). *Analysis of Incomplete Multivariate Data.* Chapman and Hall/CRC, Boca Raton, FL.

Shapley, L. S. (1953). A Value for n-Person Games. In Kuhn, H. W. and Tucker, A. W., editors, *Contributions to the Theory of Games II*, pages 307–317. Princeton University Press, Princeton.

Sheather, S. (2009). *A Modern Approach to Regression with R.* Springer Texts in Statistics. Springer, New York, NY.

Shmueli, G. (2010). To explain or to predict? *Statistical Science*, 25:289–310.

Shrikumar, A., Greenside, P., and Kundaje, A. (2017). Learning important features through propagating activation differences. In Precup, D. and Teh, Y. W., editors, *ICML*, volume 70, pages 3145–3153. Proceedings of Machine Learning Research.

Simonyan, K., Vedaldi, A., and Zisserman, A. (2014). Deep Inside Convolutional Networks: Visualising Image Classification Models and Saliency Maps. In Bengio, Y. and LeCun, Y., editors, *ICLR (Workshop Poster)*.

Simonyan, K. and Zisserman, A. (2015). Very deep convolutional networks for large-scale image recognition. In *International Conference on Learning Representations*, San Diego, CA. ICLR 2015.

Sing, T., Sander, O., Beerenwinkel, N., and Lengauer, T. (2005). ROCR: visualizing classifier performance in R. *Bioinformatics*, 21(20):7881.

Sokolva, M. and Lapalme, G. (2009). A systematic analysis of performance measures for classification tasks. *Information Processing and Management*, 45:427–437.

Staniak, M., Biecek, P., Igras, K., and Gosiewska, A. (2019). *localModel: LIME-Based Explanations with Interpretable Inputs Based on Ceteris Paribus Profiles*. R package version 0.3.11.

Steyerberg, E. (2019). *Clinical Prediction Models. A Practical Approach to Development, Validation, and Updating (2nd Ed.)*. Springer, Cham, Switzerland.

Steyerberg, E., Vickers, A., Cook, N., Gerds, T., Gonen, M., Obuchowski, N., Pencina, M., and Kattan, M. (2010). Assessing the performance of prediction models: a framework for traditional and novel measures. *Epidemiology*, 21:128–138.

Štrumbelj, E. and Kononenko, I. (2014). Explaining prediction models and individual predictions with feature contributions. *Knowledge and Information Systems*, 41(3):647–665.

Sutskever, I., Vinyals, O., and Le, Q. V. (2014). Sequence to sequence learning with neural networks. In Ghahramani, Z., Welling, M., Cortes, C., Lawrence, N. D., and Weinberger, K. Q., editors, *NIPS*, pages 3104–3112.

Tibshirani, R. (1994). Regression shrinkage and selection via the lasso. *Journal of the Royal Statistical Society, Series B*, 58:267–288.

Todeschini, R. (2010). *Useful and unuseful summaries of regression models*.

Tsoumakas, G., Katakis, I., and Vlahavas, I. (2010). Mining multi-label data. In *Data Mining and Knowledge Discovery Handbook*, pages 667–685. Springer, Boston, MA.

Tufte, E. R. (1986). *The Visual Display of Quantitative Information.* Graphics Press, Cheshire, CT, USA.

Tukey, J. W. (1977). *Exploratory Data Analysis.* Addison-Wesley, Boston, MA.

van Buuren, S. (2012). *Flexible Imputation of Missing Data.* Chapman and Hall/CRC, Boca Raton, FL.

van Houwelingen, H.C. (2000). Validation, calibration, revision and combination of prognostic survival models. *Statistics in Medicine,* 19:3401–3415.

van Rossum, G. and Drake, F. L. (2009). *Python 3 Reference Manual.* CreateSpace, Scotts Valley, CA.

Venables, W. N. and Ripley, B. D. (2002). *Modern Applied Statistics with S (4th Ed.).* Springer, New York, NY.

Štrumbelj, E. and Kononenko, I. (2010). An efficient explanation of individual classifications using game theory. *Journal of Machine Learning Research,* 11:1–18.

Wes, M. (2012). *Python for Data Analysis (1st Ed.).* O'Reilly Media, Inc.

Wickham, H. (2009). *ggplot2: Elegant Graphics for Data Analysis.* Springer-Verlag New York.

Wickham, H. and Grolemund, G. (2017). *R for Data Science: Import, Tidy, Transform, Visualize, and Model Data (1st Ed.).* O'Reilly Media, Inc.

Wikipedia (2019). *CRISP DM: Cross-industry standard process for data mining.*

Wright, M. N. and Ziegler, A. (2017). ranger: A fast implementation of random forests for high dimensional data in C++ and R. *Journal of Statistical Software,* 77(1):1–17.

Xie, Y. (2018). *bookdown: Authoring Books and Technical Documents with R Markdown.* R package version 0.7.

Index